Challenges and Applications for Hand Gesture Recognition

Lalit Kane
School of Computer Science, University of Petroleum and Energy Studies, India

Bhupesh Kumar Dewangan
School of Engineering, Department of Computer Science and Engineering, O.P. Jindal University, Raigarh, India

Tanupriya Choudhury
School of Computer Science, University of Petroleum and Energy Studies, India

A volume in the Advances in Computational Intelligence and Robotics (ACIR) Book Series

Published in the United States of America by
    IGI Global
    Engineering Science Reference (an imprint of IGI Global)
    701 E. Chocolate Avenue
    Hershey PA, USA 17033
    Tel: 717-533-8845
    Fax: 717-533-8661
    E-mail: cust@igi-global.com
    Web site: http://www.igi-global.com

Library of Congress Cataloging-in-Publication Data

Names: Dewangan, Bhupesh, 1983- editor. | Kane, Lalit, 1974- editor. |
   Choudhury, Tanupriya, editor.
Title: Challenges and applications for hand gesture recognition / Bhupesh
   Dewangan, Lalit Kane, and Tanupriya Choudhury, editors.
Description: Hershey, PA : Engineering Science Reference, [2022] | Includes
   bibliographical references and index. | Summary: "This book highlights
   various aspects of the state-of-art practices and new directions of Hand
   Gesture Recognition (HGR) research offerings"-- Provided by publisher.
Identifiers: LCCN 2021044507 (print) | LCCN 2021044508 (ebook) | ISBN
   9781799894346 (h/c) | ISBN 9781799894353 (s/c) | ISBN 9781799894360
   (ebook)
Subjects: LCSH: Gesture recognition (Computer science)
Classification: LCC TA1652 .C49 2022 (print) | LCC TA1652 (ebook) | DDC
   006.4--dc23/eng/20211104
LC record available at https://lccn.loc.gov/2021044507
LC ebook record available at https://lccn.loc.gov/2021044508

This book is published in the IGI Global book series Advances in Computational Intelligence and Robotics (ACIR) (ISSN: 2327-0411; eISSN: 2327-042X)

British Cataloguing in Publication Data
A Cataloguing in Publication record for this book is available from the British Library.

All work contributed to this book is new, previously-unpublished material.
The views expressed in this book are those of the authors, but not necessarily of the publisher.

For electronic access to this publication, please contact: eresources@igi-global.com.

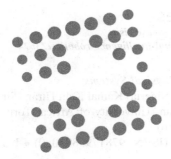

Advances in Computational Intelligence and Robotics (ACIR) Book Series

ISSN:2327-0411
EISSN:2327-042X

Editor-in-Chief: Ivan Giannoccaro, University of Salento, Italy

MISSION

While intelligence is traditionally a term applied to humans and human cognition, technology has progressed in such a way to allow for the development of intelligent systems able to simulate many human traits. With this new era of simulated and artificial intelligence, much research is needed in order to continue to advance the field and also to evaluate the ethical and societal concerns of the existence of artificial life and machine learning.

The **Advances in Computational Intelligence and Robotics (ACIR) Book Series** encourages scholarly discourse on all topics pertaining to evolutionary computing, artificial life, computational intelligence, machine learning, and robotics. ACIR presents the latest research being conducted on diverse topics in intelligence technologies with the goal of advancing knowledge and applications in this rapidly evolving field.

COVERAGE

- Fuzzy Systems
- Adaptive and Complex Systems
- Artificial Life
- Intelligent Control
- Robotics
- Natural Language Processing
- Computational Logic
- Cognitive Informatics
- Pattern Recognition
- Algorithmic Learning

IGI Global is currently accepting manuscripts for publication within this series. To submit a proposal for a volume in this series, please contact our Acquisition Editors at Acquisitions@igi-global.com or visit: http://www.igi-global.com/publish/.

Titles in this Series

For a list of additional titles in this series, please visit:
http://www.igi-global.com/book-series/advances-computational-intelligence-robotics/73674

Real-Time Applications of Machine Learning in Cyber-Physical Systems

Balamurugan Easwaran (University of Africa, Toru-Orua, Nigeria) Kamal Kant Hiran (Sir Padampat Singhania University, India) Sangeetha Krishnan (University of Africa, Toru-Orua, Nigeria) and Ruchi Doshi (Azteca University, Mexico)

Engineering Science Reference • © 2022 • 307pp • H/C (ISBN: 9781799893080) • US $245.00

Handbook of Research on New Investigations in Artificial Life, AI, and Machine Learning

Maki K. Habib (The American University in Cairo, Egypt)

Engineering Science Reference • © 2022 • 565pp • H/C (ISBN: 9781799886860) • US $345.00

Biomedical and Business Applications Using Artificial Neural Networks and Machine Learning

Richard S. Segall (Arkansas State University, USA) and Gao Niu (Bryant University, USA)

Engineering Science Reference • © 2022 • 394pp • H/C (ISBN: 9781799884552) • US $245.00

Integrating AI in IoT Analytics on the Cloud for Healthcare Applications

D. Jeya Mala (School of CS&IT, Jain University, Bangalore, India)

Engineering Science Reference • © 2022 • 312pp • H/C (ISBN: 9781799891321) • US $245.00

Applications of Artificial Intelligence in Additive Manufacturing

Sachin Salunkhe (Vel Tech Rangarajan Dr.Sagunthala R&D Institute of Science and Technology, India) Hussein Mohammed Abdel Moneam Hussein (Helwan University, Egypt) and J. Paulo Davim (University of Aveiro, Portugal)

For an entire list of titles in this series, please visit:
http://www.igi-global.com/book-series/advances-computational-intelligence-robotics/73674

701 East Chocolate Avenue, Hershey, PA 17033, USA
Tel: 717-533-8845 x100 • Fax: 717-533-8661
E-Mail: cust@igi-global.com • www.igi-global.com

Table of Contents

of Computer Science and Engineering, O.P. Jindal University, Raigarh, India

Detailed Table of Contents

Chapter 1

*Mangal Singh, Symbiosis Institute of Technology, Symbiosis
International University (Deemed), Pune, India*
*Ishan Vijay Tewari, Institute of Technology, Nirma University,
Ahmedabad, India*
*Labdhi Sheth, Institute of Technology, Nirma University, Ahmedabad,
India*

People have been continuously mesmerized by the sci-fi notion of controlling things based on hand gestures. However, only a few ponder upon the working of these gesture-based systems. This chapter explains how these gesture-based systems generally use skin-colour-based hand segmentation techniques. This chapter sheds light on and explains various colour models, various methods to use these colour models and perform skin-colour-based hand segmentation, various real-life applications of segmenting areas using skin-colour with multiple examples of hand gesture recognition, as well as other applications where skin-colour based segmentation is used.

Chapter 2

*Bhupesh Kumar Dewangan, School of Engineering, Department
of Computer Science and Engineering, O.P. Jindal University,
Raigarh, India*
Princy Mishra, Rungta College of Engineering and Technology, India

A hand gesture recognition system allows for nonverbal communication that is natural, inventive, and modern. It can be used in a variety of applications involving human-computer interaction. Sign language is a type of communication in which people communicate using their gestures. The identification of hand gestures has become one of the most important aspects of human-computer interaction (HCI). The goal of HCI is to get to a point where human-computer interactions are as natural as human-human interactions, and adding gestures into HCI is an essential research topic in that direction. The history of hand gesture recognition is initially presented in this chapter, followed by a list of technical challenges. Researchers from all over the world are working on developing a reliable and efficient gesture recognition system, particularly a hand gesture recognition system, for a variety of applications. In this chapter, the core and advanced application areas and various applications that use hand gestures for efficient interaction have been addressed.

Chapter 3

Hitesh Kumar Sharma, University of Petroleum and Energy Studies, India

Tanupriya Choudhury, School of Computer Science, University of Petroleum and Energy Studies, India

Hand gesture recognition (HGR) is a natural kind of human-machine interaction that has been used in a variety of settings. In this chapter, the authors address work in the field of HGR applications in industrial robots, with an emphasis on processing stages and approaches in gesture-based human-robot interaction (HRI), which may be beneficial to other researchers. They examine several similar research in the field of HGR that use a variety of techniques, including sensor-based and vision-based approaches. They discovered that the 3D vision-based HGR technique is a hard but interesting study field after comparing the two methodologies. Hand gesture recognition has a major application for disabled peoples. The person who is not able to speak can use these kinds of systems to control the devices or various applications in his/her machine. This technology also has vast potential in the gaming industry. The gaming industry has already adapted this technique.

Chapter 4

Bikram Pratim Bhuyan, School of Computer Science, University of Petroleum and Energy Studies, India

Shelly Garg, School of Computer Science, University of Petroleum and Energy Studies, India

Recognition of gestures has gained a lot of popularity in recent years and in the world-wide community as it has played an important role in many of the practical applications. A few of the popular practical applications are reading sign language, authentication of user, assistance provided in day-to-day life. Just to take an example, humans interact with small computers like mobile, wearable smart devices, popularly known as internet of things (IoT), which have a limitation to be embedded with in screens or keyboards. As per statistics, 26% of the population is under the age of 15 whereas some 9% of the population is above the age of 65 where they have problems performing a few activities on a daily basis, and hand gesture recognition can prove to be a good solution.

Chapter 5

Bhavana Siddineni, SRM University-AP, India
Manikandan V. M., SRM University-AP, India

Sign language has been used for a long time by deaf and mute people to communicate their thoughts and feelings. Since there is no universal sign language, the needy people use country-specific sign languages. For example, American Sign Language (ASL) is popularly used by Americans and Indian Sign Language (ISL) is commonly practised in India. Communication between two people who know the specific sign language is quite easy. But, if a mute person wants to communicate with another person who is not familiar with sign language, it is a difficult task, and a sign language interpreter is required to translate the signs. This issue motivated the computer scientist to work on automated sign language recognition systems that are capable of recognizing the signs from specific sign languages and converting them into text information or audio so that the common people can understand it easily. This chapter will be a useful reference for the researchers who are planning to start their research study in the domain of sign language recognition.

Chapter 6

Aradhana Kumari Singh, School of Computer Science, University of Petroleum and Energy Studies, India

In this chapter, hidden Markov model (HMM) is used to apply for gesture recognition. The study comprises the design, implementation, and experimentation of a system for making gestures, training HMMs, and identifying gestures with HMMs in order to better understand the behaviour of hidden Markov models (HMMs). One person extends his flattened, vertical palm toward the other, as though to reassure the other that his hands are safe. The other individual smiles and responds in kind. This wave gesture has been associated with friendship from childhood. Human motions can be thought of as a pattern recognition challenge. A computer can deduce the sender's message and reply appropriately if it can detect and recognise a set of gestures. By creating and implementing a gesture detection system based on the semi-continuous hidden Markov model, this chapter aims to bridge the visual communication gap between computers and humans.

Chapter 7

Kamal Kant Verma, Uttarakhand Technical University, India
Brij Mohan Singh, College of Engineering, Roorkee, India

RGBD-based activity recognition is quite an interesting task in computer vision. Inspired by the exemplary results obtained from automatic features learning from RGBD data, in this work a six-stream CNN fusion approach has been addressed, which is developed on 2D-convolution neural network (2DCNN) and spatial-temporal 3D-convolution neural networks (ST3DCNN). The proposed approach has six streams and runs in parallel, where the first and second streams are used to extract space and time features with the help of a ST3DCNN model. Similarly, the remaining four streams have been used to extract the temporal features by means of two motion templates on motion history image (MHI) and motion energy image (MEI) via a 2DCNN. Further, a support vector machine (SVM) is employed to generate the score from each stream. Finally, a decision level fusion scheme particularly a weighted product model (WPM) to fuse the scores is obtained from all the streams. The effectiveness of the proposed approach has been tested on popular benchmark public datasets, namely UTD-MHAD, and gives promising results.

Chapter 8

Rinki Gupta, Amity University, Noida, India

Sign language predominantly involves the use of various hand postures and hand motions to enable visual communication. However, signing is mostly unfamiliar and not understood by normal hearing people due to which signers often rely on sign language interpreters. In this chapter, a novel multi-input deep learning model is proposed for end-to-end recognition of 50 common signs from Indian Sign Language (ISL). The ISL dataset is developed using multiple wearable sensors on the dominant hand that can record surface electromyogram, tri-axial accelerometer, and tri-axial gyroscope data. Multi-channel data from these three modalities is processed in a multi-input deep neural network with stacked convolutional neural network (CNN) and long short-term memory (LSTM) layers. The performance of the proposed multi-input CNN-LSTM model is compared with the traditional single-input approach in terms of quantitative performance measures. The multi-input approach yields around 5% improvement in classification accuracy over the traditional single-input approach.

Chapter 9
Shamik Tiwari, University of Petroleum and Energy Studies, India

Deaf and hard-of-hearing persons practice sign language to converse with one other and with others in their community. Even though innovative and reachable technology is evolving to assist persons with hearing impairments, there is more scope of effort to be achieved. Computer vision applications with machine learning procedures could benefit such persons even more by allowing them to converse more effectively. That is precisely what this chapter attempts to do. The authors have suggested a MobileConvNet model that could recognise hand gestures in American Sign Language. MobileConvNet is a streamlined architecture that constructs lightweight deep convolutional neural networks using depthwise separable convolutions and provides an efficient model for mobile and embedded vision applications. The difficulties and limitations of sign language recognition are also discussed. Overall, it is intended that the chapter will give readers a thorough overview of the topic of sign language recognition as well as aid future research in this area.

Chapter 10
Hitesh Kumar Sharma, University of Petroleum and Energy Studies,
India
Tanupriya Choudhury, School of Computer Science, University of
Petroleum and Energy Studies, India

Hand gestures, as the name suggests, are different gestures made by the use of hands. Historically, hand gestures have been created to communicate with people who were unable to speak, but as new technologies have emerged, hand gestures have been used widely in different fields such as medicine, defense, IT industry. Hand gestures are being used to create TVs without remotes. Face recognition is also used to verify the user and to change channels, increase/decrease volume. To switch on or off the lights in the house, hand gesture recognition devices are being developed. Hand gesture recognition (HGR) is a natural kind of human-machine interaction that has been used in a variety of settings. In this chapter, the authors have described the application of HGR in various sectors. They have also explained the tools and techniques used for HGR.

Chapter 11

Keshav Garg, School of Computer Science, University of Petroleum and Energy Studies, India

Prabhjot Singh, School of Computer Science, University of Petroleum and Energy Studies, India

Bhupesh Kumar Dewangan, School of Engineering, Department of Computer Science and Engineering, O.P. Jindal University, Raigarh, India

Although great progress has been made by leveraging the use of sensors in various fields like human body tracking, robust, accurate, and efficient hand gesture recognition still remains an open problem. The authors draw a comparison on how the HSV (hue, saturation, value) model used in hand gesture recognition is better than the other models. Also, the focus will be laid on the different segmentation techniques that are being used nowadays and how thresholding is an important part of it. The extraction of noisy hand contours is really challenging, and for the same reason, the authors propose the usage of convexity defects to record deviation from contours and the dilution and erosion of the binary images to recognize hand gestures like rock-paper-scissors. This model has been tried and tested on different devices with various hand gestures and yields a positive result.

Preface

Automated Hand Gesture Recognition (HGR) is no more an unknown research topic nowadays. The problem is yet unresolved due to the inherent multimodality and multidimensionality. Towards shaping a real-time HGR system, most of the current research circumvents around machine learning based approaches that rigorously rinse gesture data in order to extract enough salience out of it. The intention is to develop an efficient and accurate scheme that can be deployed in real-time applications. However, the trade-off between efficiency and accuracy is not yet at an acceptable level so that it can suffice to deployment of the developed scheme. As per the current trends deep learning networks are most sought after schemes to extract the gesture data salience. There is onus on these schemes to extract orthogonality in the underlying gesture data. That means they are responsible for extracting fitting features and their classification on given gesture data. However, the HGR scenario a few years back was based on hand-crafted features, deterministic schemes, and shallow learning (as contrary to "deep-learning") classification.

An HGR system would essentially require a trained model. Once the model is in hand, it can be deployed in real-time system pipeline. The first phase in the pipeline would spot the object of interest (hands). Therefore, an effective detection and tracking of hands of continuous gesture signals would be the primary challenge. While a number of sensor-based HGR works utilizing accelerometer, gyroscope, and electromyography have been reported, newer sensors embedded in hand gloves, rings, and hand strips are constantly being introduced. When it comes to vision-based acquisition of gestural input signals, effective extraction of hand objects from the complex backgrounds becomes a complex task. With the emergence of low cost depth sensing devices, such as Kinect and XtionPro, the task of background-foreground separation has eased a bit and a number of research works have utilized or are utilizing the depth information. However, in the pre depth-sensor era, two-dimensional cameras of low to high resolution ranges were the only available options for low-cost HGR solutions. In the gesture streams captured through normal RGB camera, skin color was one potential clue to spot the hand objects. However, the skin color range suggested for human hand objects is never optimal. Especially when objects in the scene other than the hands are similar in color to the human skin, separation of the hands from the false positives goes too hard. Other hand detection clues lie in gray-level intensities and a number of schemes ranging from the naïve background subtraction to the sophisticated Kalman-filter variations do exist. The inherent procedural complexities of these schemes make them impractical for deployment.

While the revelation of the practical deployment is rare, the literature is full of approaches stating the trained model preparation. The coverage spans from elementary machine learning schemes such as nearest-neighbors and Bayesian classifiers to the sophisticated ones like deep convolutional neural networks.

The subsequent processing is dependent on the underlying HGR application. A hand gesture can be characterized by postures, hand movement paths (trajectories), and posture sequences. An HGR application may require one or more of these traits. For instance, an air-writing and recognition application may not indulge in the hand postures intricacies. It would rather capitalize on hand motion trajectories. A fingerspelling recognition application on the other hand would definitely be interested in postural details. Several fingerspelling applications would require to track both the hands as a signer may use both of her hands to articulate a fingerspell. Usually, fingerspelling is considered as a subset of Sign Language where some dedicated, one or two-handed postures denote language specific alphabets or numerals. We intentionally restrict the discussion here to manual signs only as this book precisely focuses on hand gestures. Non-manual gestures coming from facial expressions, eye brows and gazes, lip movements, etc. would lead to further recognition complexities. A full-fledged sign language recognition (SLR) system deals with all the traits; static postures, posture sequences, and interleaved trajectories. Hand postures can be analyzed for shape and texture. Once the hand region is spotted in a frame, locating the palm region and elimination of forearm sections are subsequent tasks. Specifically, the techniques interested in hand regional properties accomplish this task, though the ones dealing with kinematic properties of the hand may refrain from palm-forearm separation. Once the hand objects are segmented out, features can be computed from them and the test pattern can be inputted to a trained model for immediate labeling.

A Word on Techniques Using Kinematic Properties of Hands

Human hand can acquire enormous shapes depending upon individual degrees of freedom of the fingers, thumb, and wrist. Various strategies including geometric structure, skeletons, and convex hull analysis have come across and their improvisations towards effective hand modelling are still in progress. If skeletal clues are employed to estimate hand geometry, faster skeleton generation algorithms would be desirable, be it a pixel discarding or morphological thinning approach. Considering the silhouettes generated by segmentation of depth data, computation of medial axis or skeletons from the estimated 3D volumes is still an open area.

Due to inherent complexity in skeletonization schemes, some HGR techniques rely on contour estimation. Especially when the segmentation meets satisfactory level, contour analysis can lead to better pose recognition. However, inevitable segmentation flaws drive the implementers to look for other heuristics. Convex hull computation and associated features, e.g., angle between palm centroid and different finger points are to name a few. Points of salience in hand region area are also tracked in several HGR works. Hand geometry plays the central role in all the above stated computations.

Feature-Based Techniques for Hand Gesture Recognition

Considering the input gesture as an image, spatial domain features and spectral domain features are applied either over the entire image or the segmented ROIs. Spatial domain features are mostly dominated by statistical, histogram-based techniques and have spanned from early Histogram of Gradients (HoG) to the more sophisticated ones, e.g., Histogram of Edge Frequencies (HoEF), and Histogram of 3D Facets (H3DF). Zernike and fast-Zernike moments further expand the tally. A heavy dominance of spectral domain features is evident on gesture images where the orthogonality within the grid is captured by different Fourier transforms variants. Gabor Wavelet transform is an example. In several works, images are not explicitly acquired through the camera but several corelated parameters (signals acquired through various body sensors) form images and subsequent feature extraction is carried out over the crafted images.

Classification Strategies for HGR

Due to the inherent state-based working, Hidden Markov Model (HMM) based schemes have predominantly been used in HGR. Customizations to the HMM are in progress until the date to suit context specific needs. However, literature is full of HGR works that have applied Bayesian and adapted Bayesian classifiers, Support Vector Machines, Back Propagation Neural Networks, etc. These classifiers are adapted as per the need of gesturing application. For instance, levels predicated by SVM or KNN can be distributed across various states of a Finite State Machine (FSM). Today, the contribution of Deep Learning networks towards achieving remarkable accuracies in different image based predictive systems is well known. State-level implementations of CNNs and Deep Learning Networks have also been reported.

Application of HGR

HGR techniques shall be deemed to replace the touchscreen interfaces in future. Some straightforward applications would be in public interaction terminals, such as bank tellers, booking counters, vending machines etc. The applications may also be characterized as per the target traits (pose, pose sequence, and/or trajectories) that are to be processed. With this said, pose recognition finds importance in a gesture controlled mouse device for computers whereas trajectory recognition might help in recognizing air-written characters or words. Other application scenarios include robotics in medical surgeries, path navigation, and healthcare. Pros and cons of different techniques and rigorous analysis of the pertinent schemes should be carried out before the implementation. Possible new avenues to deploy HGR systems may find in VR/AR based games and simulations.

ORGANIZATION OF THE BOOK

The book is organized into 11 chapters. The content organization follows various tasks as per HGR pipeline and the research works centered on trained HGR model preparation. As per our study, the research on the latter is widely available. A brief description on each of these chapters is as follows:

Chapter 1 forms a good primer on color-based segmentation schemes for object of interest extraction in HGR. It thoroughly explores the usage of skin-color based hand segmentation in various HGR works and gives an insight into working, evaluation details, and comparative analysis of various color segmentation. Recommended methods, their distinctions from one another, and respective pros and cons of their applications have been discussed. The chapter recommends classifiers designed on parametric skin modeling for limited datasets. It establishes that luminance component though less useful for skin-non-skin discrimination, can be fruitful for generalizing sparse training data. It also confirms that the skin modeling approaches hold their importance while considering the evaluation of various color spaces.

Chapter 2 explores HGR from a human-computer Interaction (HCI) perspective with a view on integration of gestures in HCI applications. It talks of several techniques that fit to the flow of a typical HGR task. An overview of depth sensors towards gesture stream acquisition is also given. It is established that a combined use of RGB and depth sensor is more productive as compared to their isolated applications. Though the effective and efficient schemes to fuse the data acquired from the two modalities, IR and visible spectrum inputs, are still sought after.

Chapter 3 reviews various steps in HGR system development as per the system pipeline. It essentially explains sensor based and camera based acquisition techniques of gesture streams, skin color based segmentation, and skeleton based gesture recognition. Finally it discusses various challenges in HGR and gives an overview of potential applications.

Chapter 4 reviews the techniques that can be helpful to integrate hand gesture recognition in IoT based devices. By this integration, the claim for ubiquitously available interactive devices manageable with hand gestural interfaces is consolidated. Elderly people, persons immobile due to serious ailments, cognitively impaired people are projected to be the utmost beneficiaries of HGR-IoT integration. The intricacies of implementation are discussed along with various machine learning schemes, e.g., hidden Markov model, finite state machines, artificial neural networks, and the recent deep learning approaches.

Chapter 5 reviews several elegant works in sign language recognition (SLR). SLR, due to its inherent multimodal complexity, is still an open research area and practical solutions are still awaited. In particular, the chapter emphasizes on Indian Sign Language (ISL). It gives a complete description on ISL accompanied by its importance, history, features and grammar. It navigates through all the essential concepts for HGR such as histograms of gradients, Wavelet transform, deep learning, HMM and Convolutional Neural Networks (CNN). In addition to it, it gives glimpses of sensor based works, e.g. those using accelerometer, gyroscopes, etc. and relevant datasets.

Chapter 6 explores the usage of Hidden Markov Model (HMM) in HGR. HMM is the most popular gesture recognition technique due to its inherent application in state-based recognition tasks, such as speech recognition and all sorts of natural language processing. The chapter analyses HMM behavior by explaining articulation, training, and recognition of gestures using HMMs. Overall, it is asserted that a gesture's temporal behaviour is modelled in HMM states and the codebook maintains the constituent salience of each gesture.

Chapter 7 demonstrates fusion of six input streams acquired from space and time features, motion history image (MHI) and motion energy image (MEI). Space and time features in RGB and depth video sequences are acquired through ST3DCNN model while those in MHI and MEI are acquired through 2DCNN. Contrary to general, feature level fusion, the proposed scheme employs a decision level scheme to fuse the scores for each stream obtained through a Support Vector Machine (SVM). The six streams pertain to RGBD data for human activity recognition. Introduced scheme is an elegant approach to automatically extract orthogonal information from input streams. This is again reiterated that a combined used of color and depth input data has intuitive gains as contrary to ones achieved by isolated modalities. UTD-MHAD dataset is used for experimental evaluations.

Chapter 8 demonstrates the classification static and dynamic gestures using one-dimensional (1D) CNN and long short-term memory (LSTM). Gesture signals are acquired through multiple wearable sensors. The presented work develops a stacked CNN-LSTM deep neural network to classify multi-channel data for fifty signs of ISL. Multichannel input classification model is compared to that with single channel input and superiority of the multichannel input classification with CNN-LSTM is conveyed.

Chapter 9 proposes an American Sign Language HGR model based on MobileConvNet neural network. MobileConvNet works on depthwise separable convolutions and provides an efficient model for mobile and embedded vision applications The gestures used are isolated, static fingerspelling ASL signs. SLR intricacies and limitations are also discussed. The proposed lightweight CNN is usable in smartphone for gestural interactions. It is also mentioned that real-time sign recognition is still a challenge. It is emphasized that ASL learning efficiency should be increased.

Chapter 10 gives a walk through different possible, demonstrated applications of HGR. Intuitively the first thought of an HGR system confines us to SLR or related applications. However, gestural interaction applications is a wide area that includes remote control gestures for televisions, gesture guided weaponry, gesture assisted surgeries and medical assistance, and the list grows further. Pros, cons, and intricacies in the development of various HGR applications is covered in the content.

Chapter 11 demonstrates the workflow of HGR by developing an application of a well-known game; Rock-Paper-Scissors. Authors also cover different aspects of HGR including intricacies and cons of models, HSV color space utilization for ROI segmentation, and other HGR applications. The game doesn't use any high-end acquisition device to capture signals. An accuracy of around 94% obtained by this work asserts that abrupt illumination conditions in the playing environment impact the efficacy of the system.

We believe that the content in this book gives an insight in HGR to its readers.

Lalit Kane
School of Computer Science, University of Petroleum and Energy Studies, India

Bhupesh Kumar Dewangan
School of Engineering, Department of Computer Science and Engineering, O.P.
Jindal University, Raigarh, India

Tanupriya Chaudhury
School of Computer Science, University of Petroleum and Energy Studies, India

Introduction

AN OVERVIEW OF HAND GESTURE RECOGNITION TECHNIQUES

Gestures are the ultimate means of communication among all the beings including humans. Restricting ourselves with humans, gestures can be realized in two forms; natural and intentionally articulated. Learning of natural gestures requires a rigorous analysis of physical and psychological characteristics of individuals, while intentional articulations follow some agreed upon forms. Intentionally articulated gestures are dominated by the manual elements which involve hand postures and motions. To emphasize the gestural intentions, nonmanual elements containing facial information, e.g., eye gaze, mouthing, and facial expressions are also incorporated (Mitra, S., & Acharya, T., 2007) (Ong & Ranganath, 2005) (Zeshan, Vasishta, & Sethna, 2005). Depending upon the inclusion of manual and/or non-manual elements, various gesturing contexts can be defined. The vision-based recognition of manual gesture traits, specifically, hand postures, their sequences and motion are of great interest.

CHARACTERIZING HAND GESTURE RECOGNITION (HGR)

At the lowest level, signer articulates a hand posture (typically, a distance away from the body) and holds it stationary for a small fraction of time. The intention could be to convey a fingerspelling sign or an emblem. At times, displaying a single posture may not suffice and a sequence of postures might constitute a complete gesture. For instance, turning an open palm into a fist posture might indicate a grab action. In yet another scenario, signer moves her hand in the air. Here, motion pattern is of significance which may convey a direction with extent of the distance or it might indicate some abstract shape. In both the cases, the receiver (which is a computer in this case) needs to continuously track the gesturing hand for understanding the intention. While observing the motion pattern (trajectory), noticing the postures and locations of the hand(s) shall be desirable in even complex gestures. A typical workflow for hand gesture recognition systems is given in Fig. 1. Implementation details for each step, specifically hand segmentation, would vary depending upon the gestural traits being targeted.

Posture Recognition

An intended hand posture, which denotes a gesture, is hold stationary for a short burst (say 3-10 frames, depending upon the gesturing speed of signer). As frames in succession are likely to contain redundant information, the task is to select a key frame in the sequence containing the posture and then to recognize the posture. Description and classification calls are not necessary at each frame in the sequence. However, the selection of a key frame might require pre-processing (segmentation and related operations) for each frame. Thus, pre-processing is the most time critical operation.

Figure 1. A typical hand gesture recognition system

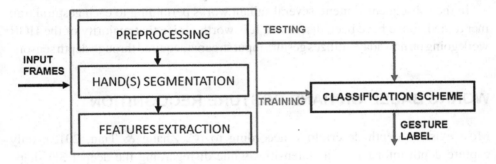

Posture Sequence Recognition

Several key postures can constitute a gesture. For instance, a hold-and-release gesture comprises of three postures; open-palm, fist, and open-palm. To handle such gestures, a window consisting of several frames can be scanned for the responses of individual posture recognitions and then a legitimate posture sequence can be detected to form a gesture. In addition, window scanning must be quick and compatible to the small multiples of temporal windows for prompt posture recognition responses.

Trajectory Recognition

Some gestures are interpreted from motion patterns of hand in the articulation space. In primary forms, association of posture information is not necessary. One example could be to draw an Arabic digit '8'. Pre-processing is essentially required in the allotted time fringe. The description and classification calls depend on gesture spotting technique employed. A general clue is to track the hand velocity and to assume the gesture boundaries at a stationary (velocity is 0) hand. Once a gesture is spotted, recognition action should be accomplished within the time fringe.

Recognition of Collaborative Traits

A merger of the traits requires simultaneous recognition of posture sequences, trajectories, and/or non-manual elements. An example is the 'Police-station' gesture from ISL which combines a posture sequence with motion pattern, while the gesture 'Girl' additionally involves non-manual element. Recognition of all the traits in a single thread of execution is complex, and therefore the solution may require parallel threads or processes of execution. Live recognition differs from that experimented with pre-recorded videos. Implemented schemes usually malfunction in live environment. Runtime parameters for live recognition vary from those in the pre-recorded videos due to variations in system parameters, resources, and noise.

In the subsequent content, several saliant works pertaining to each gestural trait mentioned above have been discussed. It is worth notice that majority of the HGR work going on nowadays utilizes gesture input streams acquired through depth sensors.

WORKS BASED ON HAND POSTURE RECOGNITION

Most existing depth descriptors, according to (C. Zhang & Tian, 2015), only capture depth information as intensities while disregarding the deeper 3D shape information. They propose the Histogram of 3D Facets (H3DF), a unique and effective descriptor for explicitly encoding 3D shape information from depth maps. The 3D local support surface is defined by a 3D Facet connected with a 3D cloud point. The proposed H3DF descriptor can successfully capture both 3D shapes and structures of various depth maps by using resilient coding and circular pooling 3D Facets from a depth map. We enhance the proposed H3DF by integrating it with an N-gram model and dynamic programming to handle the recognition challenges of dynamic actions and gestures. Two public 3D static hand gesture datasets, one dynamic hand gesture dataset, and one popular 3D action recognition dataset are used to test the proposed descriptor. The results of the recognition outperform or are comparable to state-of-the-art outcomes.

A novel local descriptor for video sequences is presented by (Klaser, Marszalek, & Schmid, 2008). Histograms of oriented 3D spatio-temporal gradients form the basis of the proposed descriptor. They make a four-fold contribution. Firstly, They design a memory-efficient approach based on integral movies to compute 3D gradients for various scales. Then they propose a regular polyhedron-based generic 3D orientation quantization. Further they evaluate all descriptor parameters in depth and optimise them for action recognition. Finally, they demonstrate that their descriptor outperforms the state-of-the-art by applying it to a variety of action data sets.

(Oreifej & Liu, 2013) propose a new descriptor for activity recognition from depth sensor videos. Previous descriptors often compute form and motion data separately, which means they frequently miss complicated joint shape motion cues at the pixel level. They characterise the depth sequence, on the other hand, with a histogram that captures the distribution of the surface normal orientation in 4D space of time, depth, and spatial coordinates. They make 4D projectors to build the histogram, which quantize the 4D space and represent the 4D normal's probable directions. They also use the vertices of a regular polychoron to start the projectors. As a result, they use a discriminative density measure to refine the projectors, causing more projectors to be generated in the directions where the 4D normals are denser and discriminative. They show that their descriptor better captures the joint shape-motion cues in the depth sequence, and consequently outperforms the state-of-the-art on all relevant benchmarks, thanks to extensive tests.

(Hernandez-Vela et al., 2012), propose a Bag-of-Visual-and-Depth-Words (BoVDW) model for gesture identification, which is an extension of the Bag-of-Visual-Words (BoVW) model and takes advantage of the multimodal fusion of visual that depth cues. In a late fusion approach, state-of-the-art RGB and depth features, including a novel suggested depth descriptor, are analysed and fused. The method is part of a continuous gesture recognition pipeline that uses the Dynamic Time Warping (DTW) algorithm to partition motions prior to recognition. In comparison to a normal BoVW model, the technique performs better in public data sets within their gesture recognition pipeline.

Works Based on Posture Sequence Recognition

(Kane & Khanna, 2019), show how a series of seemingly random hand postures can yield significant dynamic gestures that can be used in computer, television, and gaming interface controllers. Select descriptors must be fast enough to meet the live recognition requirements in order to construct deployable systems with these gestures. In their framework for a practical system that can recognise continuous dynamic gestures with short-duration posture sequences. To define hand silhouettes, a depth-based modification to the form matrix is designed, which provides a speedier alternative to region-based descriptors. The 1-nearest neighbour approach and a depth matrix are used to recognise postures. A dynamic naive Bayes classifier, in conjunction with an adaptive windowing technique, predicts posture sequence labels.

(Macdorman & Iwahori, 2014), Continuous hand gesture recognition is driven by applications that require the natural usage of the human hand as a human–computer interface. Gesture segmentation is used to determine the beginning and end points of important motions while ignoring unintended movements in gesture recognition. Unfortunately, due to unrestricted spatiotemporal fluctuations in gestures and the co-articulation and movement epenthesis of subsequent gestures, gesture segmentation remains a tough issue. Errors in hand picture segmentation also lead the estimated hand motion trajectory to differ from the actual one. To discern important gestures from unintended motions, they employ gesture spotting. They propose alternatively using a novel set of features: the (a) orientation and (b) length of an ellipse least-squares fitted to motion-trajectory points and (c) the location of the hand, to eliminate the impacts of variations in a gesture's motion chain code (MCC). The attributes are intended to assist in the classification of conditional random fields. The recognition rate for isolated gestures was only 69.6% using the MCC as a feature vector, but it increased to 96.0 percent utilising the proposed features, a 26.1 percent improvement. The proposed features were recognised at an 88.9% rate for continuous gestures. The efficacy of the proposed strategy is demonstrated by these outcomes.

WORKS BASED ON TRAJECTORY RECOGNITION

Towards the recognition of mid-air writing trajectories, many researchers have proposed novel approaches. In a Kinect-based trajectory recognition work, (Stern, Shmueli, & Berman, 2013) target the recognition of dynamic gestures using representative sub-segments of the gesture, referred to as most discriminating segments (MDSs). Rather than collecting and identifying the whole motions, a more discriminative classifier can be created by automatically extracting and recognising tiny representative portions. A MDS is a gesture sub-segment that is the most distinct from all other gesture sub-segments. The MDSLCS algorithm recognises MDSs using a modified longest common subsequence (LCS) measure to classify gestures. Adaptive window settings, which are influenced by the sequential outcomes of numerous calls to the LCS classifier, are used to extract MDSs from a data stream. Gestures with substantial motion variations are replaced by many types of lesser variation throughout the pre-processing stage. Learning of these shapes by reusing the LCS to determine similarity between gesture trajectories after adaptive clustering of a training set of gestures. When evaluated using a set of pre-cut free hand digit (0–9) motions, the MDSLCS classifier had a recognition rate of 92.6 percent, whereas hidden Markov models (HMMs) had an accuracy of 89.5 percent.

(Nyirarugira & Kim, 2015) present a stratified gesture identification approach for natural human–computer interaction that combines rough set theory with the longest common sub sequence method to identify free-air motions. Gesture vocabulary are frequently made up of gestures that are closely related or involve motions that are part of a larger group. If no other steps are taken, the accuracy of most classifiers will be reduced. In their research, they discovered that motions are encoded in orientation segments, which makes analysis easier and reduces processing time. They create rough set decision tables conditioned on the longest common sub sequences to improve the accuracy of gesture identification on ambiguous gestures; the decision tables store discriminative information on ambiguous gestures. They efficiently perform stratified gesture recognition in two steps: first, a gesture is classified in its equivalence class using the normalised longest common subsequence paired with rough set decision tables, and then it is recognised using the normalised longest common subsequence paired with rough set decision tables. The recognition rate of the longest common subsequence has improved in both preisolated and stream gestures, according to the findings of the experiments.

(Frolova, Stern, & Berman, 2013) offer a method for trajectory classification that can be used to recognise dynamic free-air hand gestures. These gestures are unrestricted and drawn in the open air. Their method is based on the LCS classification algorithm, which uses the longest common subsequence (LCS). A learning pre-processing stage is used to produce a probabilistic 2-D template for each gesture, allowing different trajectory distortions with varying probabilities to be taken into consideration. The most likely LCS (MPLCS) is a modified LCS that is used to assess the similarity between the probabilistic template and the hand gesture sample. The length and probability of the retrieved subsequence are used to make the final decision. Validation tests employing a cohort of gesture digits from video-based capture demonstrate that the method is promising, with a recognition rate of over 98 percent for preisolated digits from video streams. To make gesture character input easier, the MPLCS algorithm can be implemented into a gesture recognition interface. This has the potential to dramatically improve the usability of such interfaces.

According to (Mukherjee, Ahmed, Dogra, Kar, & Roy, 2019), air-writing is the practise of writing characters or words in free space without the use of a hand-held device, utilising finger or hand movements. They use web-cam video as input in their study to address the problem of mid-air finger writing. Despite recent developments in object identification and tracking, accurate and reliable detection and tracking of the fingertip remains a difficult task, owing to the fingertip's small size. Furthermore, the lack of a consistent delimiting criterion makes the commencement and termination of mid-air finger writing difficult. To address these issues, they offer a new writing hand posture recognition technique based on geometrical features of the hand for initialization of air-writing using the Faster R-CNN framework for accurate hand detection, hand segmentation, and lastly counting the number of lifted fingers. They also present a method for detecting and tracking fingertips that is based on a new signature function termed distance-weighted curvature entropy. Finally, to identify the end of the air-writing motion, a fingertip velocity-based termination criterion is used as a delimiter. Experiments reveal that the suggested fingertip identification and tracking algorithm outperforms existing methods, with an average precision of 73.1 percent. Character recognition trials show that the suggested air-writing system has a mean accuracy of 96.11 percent, which is equivalent to existing handwritten character recognition systems.

Because there is no user-side interface device, air-writing is an appealing technique of human-machine interaction, according to (Mohammadi & Maleki, 2019), It can be employed in a variety of applications in the future after removing existing limits and addressing current obstacles. They presented an air-writing system that uses Kinect depth and colour images to detect single characters such as digits or letters, as well as connected characters such as numbers or phrases. Automatic clustering, slope variations detection, and a novel analytical classification are proposed as new approaches for removing noise in the trajectory from the depth image and hand segmentation, extracting the feature vector, and identifying the character from the feature vector, respectively, in this system. The suggested system can correctly detect single and connected characters with an average recognition rate of 97 percent, according to experimental results. Because a novel analytical classifier is used in the suggested system, the character recognition time is very short, around 3 ms. The suggested classifier outperforms the SVM, HMM, and K-nearest neighbours classifiers in terms of speed and precision in a comparison of four classifiers.

With proven track record on other machine learning based tasks, Convolutional Neural Networks (CNNs) have been experimented in air-writing recognition too. Several works based on non-temporal trajectory sequences, such as Chinese calligraphy (J. Zhang, Guo, & Fan, 2019), give potential clues for the recognition of air-writing gestures. Chinese calligraphy, a priceless cultural asset, is well-liked by many individuals, and their objective is to pursue calligraphy and make technical contributions to the industry. The automatic recognition of calligraphy styles using image processing techniques has crucial implications in art collections and auctions, among other things. Traditional feature operators have a number of flaws that make contemporary methods like convolutional neural networks more appealing (CNN). However, the majority of research concentrate on the classification of five fundamental fonts, which are distinct in articulation styles. Four types of standard font styles are classified using a novel CNN structure in which two squeeze-and-excitation modules that emphasise informative feature maps while suppressing useless features are embedded after convolution layers, and a Haar transform layer that fuses the features is imposed before the softmax layer. The relevance of the proposed structure above other networks in both font and style categorization is demonstrated by the results of the experiment.

CONCLUSION

This chapter has given an overview on three principle HGR traits; hand posture, posture sequences, and hand motion trajectories and discussed the relevant techniques to each trait. As a global observation, it appears that the works developed so far though heading towards practical implementation but a concrete HGR system is still farsighted. In air-writing, descriptors such as MDSLCS, Rough set theory, and Equi-Polar Signature (EPS) (Kane & Khanna, 2017) (Kane & Khanna, 2016) seem promising. but the computational cost incurred requires further rectification in the schemes. Recently, CNN based deep learning networks have also shown comprehensive results. Primary trajectory recognition applications do not require to track hand postures and need to track hand reference point only. A comparatively dauting task is the tracking of postures which is otherwise hindered by out-of-plane view variations in articulations. Still, H3DF, HON4V and BoVDW feature based works have demonstrated satisfactory results on open datasets.

Lalit Kane
School of Computer Science, University of Petroleum and Energy Studies, India

Bhupesh Kumar Dewangan
School of Engineering, Department of Computer Science and Engineering, O.P. Jindal University, Raigarh, India

Tanupriya Choudhury
School of Computer Science, University of Petroleum and Energy Studies, India

REFERENCES

Frolova, D., Stern, H., & Berman, S. (2013). *Most Probable Longest Common Subsequence for Recognition of Gesture Character Input*. Academic Press.

Hernandez-Vela, A., Bautista, M. A., Perez-Sala, X., Ponce, V., Baro, X., Pujol, O., . . . Escalera, S. (2012). BoVDW: Bag-of-Visual-and-Depth-Words for gesture recognition. *Proceedings - International Conference on Pattern Recognition*, 449–452.

Kane, L., & Khanna, P. (2016). A Framework to Plot and Recognize Hand Motion Trajectories towards Development of Non-tactile Interfaces. *Procedia Computer Science*, *84*, 6–13. doi:10.1016/j.procs.2016.04.059

Kane, L., & Khanna, P. (2017). Vision-Based Mid-Air Unistroke Character Input Using Polar Signatures. *IEEE Transactions on Human-Machine Systems*, *47*(6), 1077–1088. doi:10.1109/THMS.2017.2706695

Kane, L., & Khanna, P. (2019). Depth matrix and adaptive Bayes classifier based dynamic hand gesture recognition. *Pattern Recognition Letters*, *120*, 24–30. doi:10.1016/j.patrec.2019.01.003

Klaser, A., Marszalek, M., & Schmid, C. (2008). A Spatio-Temporal Descriptor Based on 3D-Gradients To cite this version : A Spatio-Temporal Descriptor Based on 3D-Gradients. *19th British Machine Vision Conference*, 275–286.

Macdorman, K. F., & Iwahori, Y. (2014). *A novel set of features for continuous hand gesture recognition.* doi:10.1007/s12193-014-0165-0

Mitra, S., & Acharya, T. (2007). Gesture recognition: A survey. *IEEE Transactions on Systems, Man and Cybernetics. Part C, Applications and Reviews*, *37*(3), 311–324. doi:10.1109/TSMCC.2007.893280

Mohammadi, S., & Maleki, R. (2019). Real-time Kinect-based air-writing system with a novel analytical classifier. *International Journal on Document Analysis and Recognition*, *22*(0123456789), 113–125. Advance online publication. doi:10.100710032-019-00321-4

Mukherjee, S., Ahmed, S. A., Dogra, D. P., Kar, S., & Roy, P. P. (2019). Fingertip detection and tracking for recognition of air-writing in videos. *Expert Systems with Applications*, *136*, 217–229. doi:10.1016/j.eswa.2019.06.034

Nyirarugira, C., & Kim, T. (2015). Signal Processing : Image Communication Stratified gesture recognition using the normalized longest common subsequence with rough sets. *Signal Processing Image Communication*, *30*, 178–189. doi:10.1016/j.image.2014.10.008

Ong, S. C. W., & Ranganath, S. (2005). Automatic sign language analysis: A survey and the future beyond lexical meaning. *IEEE Transactions on Pattern Analysis and Machine Intelligence*, *27*(6), 873–891. doi:10.1109/TPAMI.2005.112 PMID:15943420

Oreifej, O., & Liu, Z. (2013). HON4D: Histogram of oriented 4D normals for activity recognition from depth sequences. *Proceedings of the IEEE Computer Society Conference on Computer Vision and Pattern Recognition*, 716–723. 10.1109/CVPR.2013.98

Stern, H., Shmueli, M., & Berman, S. (2013). Most discriminating segment – Longest common subsequence (MDSLCS) algorithm for dynamic hand gesture classification. *Pattern Recognition Letters*, *34*(15), 1980–1989. doi:10.1016/j.patrec.2013.02.007

Zeshan, U., Vasishta, M., & Sethna, M. (2005). Implementation of Indian Sign Language in educational settings. *Asia Pacific Disability Rehabilitation Journal*, *16*(1), 16–40.

Zhang, C., & Tian, Y. (2015). Histogram of 3D Facets: A depth descriptor for human action and hand gesture recognition. *Computer Vision and Image Understanding*, *139*, 29–39. doi:10.1016/j.cviu.2015.05.010

Zhang, J., Guo, M., & Fan, J. (2019). A novel CNN structure for fine-grained classification of Chinese calligraphy styles. *International Journal on Document Analysis and Recognition*, *22*(2), 177–188. doi:10.100710032-019-00324-1

Chapter 1
Skin–Colour–Based Hand Segmentation Techniques

Mangal Singh
Symbiosis Institute of Technology, Symbiosis International University (Deemed), Pune, India

Ishan Vijay Tewari
Institute of Technology, Nirma University, Ahmedabad, India

Labdhi Sheth
Institute of Technology, Nirma University, Ahmedabad, India

ABSTRACT

People have been continuously mesmerized by the sci-fi notion of controlling things based on hand gestures. However, only a few ponder upon the working of these gesture-based systems. This chapter explains how these gesture-based systems generally use skin-colour-based hand segmentation techniques. This chapter sheds light on and explains various colour models, various methods to use these colour models and perform skin-colour-based hand segmentation, various real-life applications of segmenting areas using skin-colour with multiple examples of hand gesture recognition, as well as other applications where skin-colour based segmentation is used.

INTRODUCTION

Machines have been learning from humans and humans understand and interpret visual information. This visual information can be our facial emotions, the motion of the lips moving while speaking, using various body parts to perform some

DOI: 10.4018/978-1-7998-9434-6.ch001

activity like running, or even using hands to gesture something. This functionality of hand gesture recognition has become the latest area of innovation in the field of Computer Science.

Hand Gesture Recognition is basically controlling a machine or some functionality of the machine by using hand gestures. Hand Segmentation is one of the steps for Hand Gesture Recognition. The applications of Hand Gesture Recognition have come into focus since Human-Computer Interaction (HCI) has become a hotspot technology. The human-computer interface (HCI), a multidisciplinary field of computer science is one of the most widely used applications of hand segmentation. Hand gestures and movement are some of the modes of communication that we humans use for communications and HCI tries to achieve this where the gestures are understandable for the computer to interpret. In recent times, hand gesture applications like writing on the screen using fingertip movement or the advanced version of it are to make a virtual keyboard and use the fingertip position to select a character or a number or a special case character, volume modulation of smart home devices, using Internet of Things (IoT)devices, a gaming joystick, the popular voice search used by various search engines, etc. have gained popularity.

Thus, for gesture recognition to be effective, hand Segmentation should be accurate and effective. Skin-based Hand Segmentation is one of the methods of hand segmentation, rather a very popular method for hand segmentation because of its, firstly high processing speed and the invariance against rotations, and secondly partial occlusions and changes in pose. It also proves to be significantly effective in day-to-day applications like computer games, robot control, and smart home devices. A major reason for it to be effective is that the color of the hands has relative stability that is, it doesn't change much, and thus can be used to distinguish from other objects for segmentation. In this chapter, the authors discuss several techniques of Skin-Based Hand Segmentation, their drawbacks, and advantages, and the future scope of this technique of hand segmentation.

SKIN-COLOR BASED HAND SEGMENTATION: BACKGROUND

First of all, let us first understand the word, "segmentation". Segmentation is the process of predicting the class of each pixel. Now since our image can be seen as a matrix representation of the pixels thus during the process of segmentation we segment the pixels as, in simpler terms, a hand or background.

Skin colour hand segmentation does a similar task where it searches for locations of human skin pixels and segments them into regions with skin and non-skin pixels. It is one of the most preferred feature-based detection techniques but it is not a simple task to perform. Human skin has a wide range of shades and thus skin segmentation is a challenging task as the imaging conditions and illumination factor can affect the total image thus making it difficult to target and extract the skin.

Skin detection models present different results as the detection technique should work dynamically with the camera distance, background colour, and illumination and thus give different results. Thus it becomes difficult to perform the segmentation accurately and make it scalable for all skin types that vary as per gender and geographical conditions.

Humans have been working on new technologies which try to imitate the powerful human brain, for instance, Deep Neural Networks which are based on and imitate the nature of the neural networks present in the human brain that process an output based on the input that is given.

Similarly, humans can recognize skin colour, and thus it has been used for segmentation applications. As we discussed earlier that skin colour hand segmentation is not a very easy task then the question rightly arises why prefer it? Well, the answer to this intriguing question is that skin colour has characteristic properties which makes it the perfect choice over the other hand segmentation techniques as it has been proven to be a more useful method for detection, localization, and tracking (Vladimir et al. 2003). Then the question arises of whether the skin colour segmentation method is robust? The answer to this question is that colour allows fast processing, and thus this property makes the segmentation technique highly robust to geometric variations.

The usefulness of this method of hand segmentation would be more clear by its following applications:

1. It is used in Image Content Filtering where we filter and manipulate the pixels of the images as per our interest. In our case, we filter the pixel data as per its intensity values to differentiate the skin pixels from the background pixels.
2. It is also used in Video Compression which has been an essential part of our daily lives now as it is responsible for fast communication of high quality and high-resolution videos which takes a larger space. Video compression generally is lossy as it removes the redundant frames but in the case of content-aware video compression, we focus on the segmentation process which helps us to extract frames with objects of interest (Nisha Devi et al. 2017).
3. Perpetual color constancy is the ability that states the color of the objects will appear the same under a variety of different illuminations. It is an ability that human eyes have. Through Image color, balancing machines try to achieve the same quality of a photo having the same color under different illuminations.

Apart from the above applications, face and hand detection and tracking uses the skin segmentation method extensively.

While performing skin color-based segmentation, one decides three factors.

1. Firstly, which color space model will fit in well.
2. Secondly, how should the skin color distribution be modeled
3. Lastly, how to process the color segmented result for detection or tracking.

We will be discussing these points in the upcoming sections.

Color Space Models

Color space models are the mathematical way of describing colors. They simply define the composition of the colors that form the image pixels. Color spaces are widely used, for instance, a photographer is concerned with transmitted light that creates the images that are viewed on the screens, and hence sRGB, Adobe RGB, and Pro-Photo RGB color space models are used by them.

In the case of a printer, the focus is more on the reflected light, and on account of this the subtractive colors which are Cyan, Magenta, and Yellow that compose the CMY color space model are used here. This model is used more often for the printing industry applications. Color space models like the HSB which stands for Hue, Saturation, and Brightness are used by the artists.

When we divide the color spectrum as a multidimensional model then it makes a color space model. For instance, let's take an example of the RGB color space model, this is a 3-dimensional color space model with coordinates for Red, Green, and Blue respectively. Let's learn about the popular color space models on which skin color-based segmentation was performed:

1. RGB color space model: This color space model is composed of three primary colors: red, green, and blue. It is a hardware-oriented model. It originated from CRT display applications. This model is not a favorable choice for color-based segmentation due to the high correlation between the channels, significant perpetual non-uniformity, and mixing of chrominance and luminance data (Vladimir et al. 2003). RGB colors make CMYK color space models which are Cyan, Magenta, Yellow, and Black. The absence of red makes cyan, the absence of green makes magenta and similarly absence of blue makes yellow. The absence of all three colors makes black and the presence of all three colors makes black.

2. Normalized RGB: This color space model is simply attained by the normalization procedure on RGB values as,

$$r = R / \left(R + G + B \right) \tag{1}$$

$$g = G / \left(R + G + B \right) \tag{2}$$

$$b = B / \left(R + G + B \right) \tag{3}$$

where, *r, g,* and *b* are called normalized components, and the sum of these components is 1. This color space model is used for the matte surfaces while ignoring the ambient light. Under certain assumptions, normalized RGB remains unchanged to changes of surface orientation relative to the light source.

3.   3. HSI model: Brightness is the achromatic notion of intensity, hue is the dominant wavelength in the mixture of light waves and saturation is the amount of white light mixed with hue. These three are the distinguishing factors of the RGB model. But since brightness is nearly impossible to measure because of its subjective nature we bring in Intensity. Intensity is the same achromatic notion that we have seen in grey-level images. Thus hue, saturation, and intensity make the HSI model. The HSI model is more on the application and perception orientation side. This model was preferred for skin-color segmentation because of the explicit discrimination between luminance and chrominance properties. However, it wasn't taken forward due to several undesirable features like hue discontinuities and the computation of "brightness" that conflicted with color vision properties. The cyclic nature of polar coordinates in the hue-saturation space makes it inconvenient for the skin color segmentation as it results in parametric skin color models where the desired are tight clusters of skin colors for optimal performance.

$$H = \arccos \frac{\frac{1}{2}\left(\left(R - G \right) + \left(R - B \right) \right)}{\sqrt{\left(\left(R - G \right)^2 + \left(R - B \right)\left(G - B \right) \right)}} \tag{4}$$

$$S = 1 - 3 \frac{\min\left(R, G, B \right)}{R + G + B} \tag{5}$$

$$V = \frac{1}{3}\left(R + G + B\right) \tag{6}$$

4. 4. TSL model: The Tint, Saturation, and Lightness space model is the normalized RGB model transformed in a more intuitive way to inculcate hue and saturation. This model was in a debate that it is superior to the other color spaces for the skin modeling and (Terrillon et al. 2000) have validated their points by comparing nin different color spaces for skin modeling with a unimodal Gaussian joint.

$$= \left[9 \,/\, 5\left(r'^2 + g'^2\right)\right]^{1/2} \tag{7}$$

$$T = \{\arctan \frac{\left(\frac{r'}{g'}\right)}{2\pi + \frac{1}{4}, g' > 0} - \arctan \frac{\left(\frac{r'}{g'}\right)}{2\pi + \frac{3}{4}, g' > 0} 0, g' = 0 \tag{8}$$

$$L = 0.299R + 0.587G + 0.114B \tag{9}$$

5. 5. YCbCr model: YCbCr is a family of color spaces with Y as the luminance component Cb is the blue difference and Cr accounts for the red difference. This model has been widely used for skin modeling due to its nature of transformation simplicity and explicit separation of luminance and chrominance. Many of the researches done on skin color-based hand segmentation have used this color model or the adaptive version of this.

$$Y = 0.299R + 0.587G + 0.114B \tag{10}$$

$$C_r = R - Y \tag{11}$$

$$C_b = B - Y \tag{12}$$

The above mentioned are the color space models that have been widely experimented with to perform skin color-based segmentation.

Choosing the right color space model seems to be a crucial task for skin-based segmentation and therefore it is very natural to ask whether which color space model will be the best choice for skin segmentation? As per the readings, it was observed that many papers didn't provide any strict choice for their color space for the skin detection, it is assumed to be because the possibility to obtain optimal results for skin detection was performed on a limited dataset with almost any color space model. On the other hand, a few papers did devote their time to comparing the outcomes of different color spaces like Vladimir et al. (2003), Zarit et al. (1999), Terrillon et al. (2000), Gomez (2000), Gomez and Morales (2002), Stern and Efros (2002).

Color space "goodness" for skin modeling is measured as; first, by training and test set classification error which is computed after the color model parameter is estimated. This metric to find the goodness of fit for skin modeling on a given dataset is a well-known classifier performance evaluator. The second measure quantifies the compactness of the skin cluster by the skin and non-skin color overlap for the given colorspace. Both of these are a family of measures to evaluate the 'in general' performance of skin detection while being independent of color modeling strategy. Although both of these measures can provide the idea of the distribution of skin to non-skin pixels the feasibility for evaluation of the colorspace goodness seemed doubtful to the authors of (Vladimir et al. 2003).

In some of the previous researches like (Shin et al. 2002), the authors had used scattered matrices of skin and non-skin clusters, and skin and non-skin histograms overlap as the color space performance metric. They believed that the RGB color space model provided the maximum separability in the skin and non-skin pixels and had said that dropping the luminance component worsens the separability significantly. Whereas the authors of (Vladimir et al. 2003) strongly believed that the reasonable colorspace comparison is for a certain skin distribution model and not for the one carried in general.

Let us now understand how crucial it is to choose a color space model with respect to the method used for skin color-based hand segmentation. Three different types of methods for skin segmentation are discussed in the next topic. One of them is parametric skin distribution modeling. The performance of parametric skin classifiers depends majorly on the choice of the color space models. The other type of method which is defining skin region explicitly is benefitted the most when an appropriate type of color space model is used and the parameters to those models are appropriately learned. The third type of method is the non-parametric model which uses the probability value and therefore is almost independent of the color space model choice. The authors of [9] have provided the theoretical results on this by stating that if $D(x)$ is the optimal skin detector in the given color space C then for the invertible colorspace transformation rule which is T: C →C' there exit a D'(x') in the invertible color space C', this in fact has the same true positive and false-positive rates.

The skin tone is guided by chrominance coordinates. On account of this several previous types of research have dropped the luminance component of the colour space model for skin detection. This decision is said to be logical too for many reasons as; firstly the skin tone is majorly affected by chrominance than luminance, secondly, the dimensionality reduction by dropping it simplifies the consequent color analysis, thirdly, skin color differs from person to person which is mostly dominated by brightness rather than tone and lastly illumination conditions heavily influences the colour of the hand in the frame and the ultimate goal is to reduce this so as to make the skin segmentation systems more robust to these changes. Some of the authors have even stated that dropping the luminance factor is only for the matter of training data as the skin segmentation classifier learns from the training data where there are a few face images under similar kind of illumination conditions and would work ahead in test data too.

Therefore, concluding the various researches and discussions done it was observed that switching to adaptive color space method for hand segmentation was preferred. Color space quality measure is the metric to determine optimal color space for a given video frame and dynamic change of color space contributes to building the robustness of the segmentation technique for tracking and detection. The experimental data says that:

- Five-color space chromaticity planes namely: normalized RGB, HS, YQ, and CrCb
- RG plane of RGB space
- HS planes

All the above planes have performed equally and have given better results than others.

METHODS FOR SKIN COLOUR-BASED HAND SEGMENTATION

Having now introduced to you what we mean by skin color-based hand segmentation, let us now understand the various methods and color space models for the same.

Skin color can be detected using pixel-based and region-based methods. Pixel-based skin detection methods classify each pixel to skin or non-skin class individually. Whereas in the case of region-based methods the spatial arrangement of the skin pixels is taken into account (Vladimir et al. 2003).

The first step is to choose an appropriate color space model which was discussed in the previous section. Our next goal is to segment between the skin and non-skin pixels. Quite a few techniques are mentioned in literature by various researchers wherein they have adopted different color spaces and various skin modeling techniques. However, for real-time tracking only a handful of them are useful. Real-time tracking challenges the researchers with its timing complexity and that it should have low false positive and false negative rates.

Skin color-based segmentation has been divided into three major categories namely region-based, parametric, and non-parametric distribution modeling methods.

Figure 1. Different techniques under skin color based segmentation

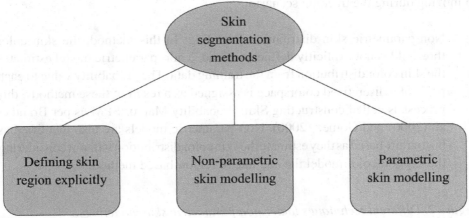

For all these methods a metric is calculated which calculates the pixel color to the skin tone. The methods along with their subdivisions are explained as below:

1. Defining skin region explicitly: This is a very simple method where the boundaries are set explicitly for a skin cluster. For instance, in the RGB colorspace model, the values of R, G, B are set in a way that ranges into the skin tone that is:

R>95 and G>40 and B>20 and (max{R, G, B} - min{R, G, B}) > 15 and |R-G|>15 and R>G and R>B (Vladimir et al. 2003).

Colors that have the same chromaticity like dark, yellowish and pale skin colors tend to follow a curve similar to Planckian locus under light sources and under different correlated color temperatures (M. Soriano et al. 2003). Planckian locus defines the points in color space that would be followed by an incandescent blackbody radiator as the temperature of the points changes. Thus a skin locus can be defined as a region that contains sample skin pixels belonging to distinct races and captured under various illumination conditions.

This method of skin segmentation is a very rapid process as the threshold value is set explicitly and then the value of each pixel is compared with that value to classify it as skin or non-skin and this should make it suitable for preprocessing images for hand gestures but there is a drawback to it. Since this method suggests explicitly choosing the skin clusters for a color space model it cannot be directly applied in real-time hand gesture recognition and tracking because of the factor that it won't be capturing the different tones, lighting conditions and can even detect all the possible skin tones ins a frame and will not be able to distinguish between different objects in moving during the tracking scenario.

2. Non-parametric skin distribution modeling: In this method, the skin color threshold is not explicitly defined, instead, a non-parametric model estimates the skin color distribution from the training data. The probability value to each point of a discretized colorspace is assigned as a result of these methods; this process is called constructing Skin Probability Map or SPM as per Brand et al. (2000) and Gomez (2000). Non-parametric models are also identified as histogram-based as they estimate the skin color distribution without considering the explicit color model the one like in region-based method.

Figure 2. Different techniques under non parametric skin modeling

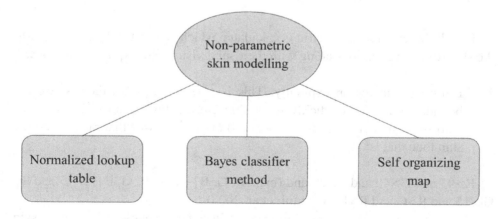

The following are the various researched nonparametric skin distribution modeling methods:

a. Normalized lookup table method: This method performs skin color-based segmentation using the histogram method. It is used in several face tracking and detection algorithms (Sigal et al. 2000), (Soriano et al. 2000), (Birchfield 1998),(Chen et al. 1995), (Zarit et al. 1999), (Schumeyer and Barner 1998). The histogram method works in a way where the color space is quantized into a number of bins. Each of these bins corresponds to a particular range of color components which further forms either a 2d or 3D histogram. This histogram which is formed is called a lookup table. Each bin stores the frequency of the occurrence of that color in the training images with skin as we are using nonparametric methods for skin modeling.

After training, the histogram counts are normalized to achieve discrete probability distribution. It is computed as:

$$Pskin(c) = skin[c] / Norm, \tag{13}$$

where skin[c] is the value of the histogram bin corresponding to the vector c and Norm is the sum of all bin values, Norm can even take the maximum bin value present. Thus the lookup table constitutes the likelihood of the color to be a skin pixel.

b. b. Bayes classifier method: Bayes classifier calculates the posterior probability using the likelihood and prior probability. The probability calculated in the above step is a conditional probability which is the probability of observing a color c given the skin pixel. Here, the probability of observing skin given concrete c value is calculated as:

$$P(skin \mid c) = P(c \mid skin)P(skin) / P(c \mid skin)P(skin) + P(c \mid non\text{-}skin)P(non\text{-}skin) \tag{14}$$

$P(c \mid non\text{-}skin)$ and $P(c \mid skin)$ are calculated as (i) and $P(skin)$ and $p(non\text{-}skin)$ is calculated by estimating the count from the training sample. A skin detection rule can be set by assigning an inequality as $P(skin \mid c) >= \theta$ where θ is some threshold value. The ROC curve which is the Receiver operating characteristics curve states the relationship between correct and false detections for the classification rule as a function of the detection rule. ROC curve says that $P(skin \mid c) >= \theta$ does not change with the choice of prior probability reason being the nature of the Bayes model. This means that prior probability $P(skin)$ only affects the choice of threshold.

Equation (14) can be simplified as:

$$P(skin \mid c) / P(non\,skin \mid c) = P(c \mid skin)P\big(skin\big) / P(c \mid non\text{-}skin)P\big(non\text{-}skin\big) \tag{15}$$

Comparing (15) with the threshold decision rule we get:

$$P(c \mid skin) / P(c \mid non\text{-}skin) > \theta \tag{16}$$

Where $\theta = Kx\big(1 - P\big(skin\big)\big) / P\big(skin\big)$

The above equations show us that the choice of prior probabilities does not affect the overall behavior of the detector instead for any prior probability it is possible to choose the appropriate K such that obtains a similar detection threshold θ.

c. Self-Organizing Map method: It is a popular variant of unsupervised artificial neural networks. SOMs are used for dimensionality reduction, they produce a low dimensional typically a 2D discrete representation of the input space of the training samples called a map. SOMs use a neighborhood function to preserve the topological properties of the input space.

This method is a transductive learning algorithm to track the human hand in the video frame. The color distribution of each time frame is approximated by the 1D self-organizing map (SOM) in which the scheme promotes growth, clipping, and blending to automatically find the appropriate number of color groups. Due to the dynamic background and changing lighting conditions, the color distribution over time may not be fixed. An algorithm for SOM transduction is proposed to learn the non-stationary color distribution in the HSI color space by combining the supervised and unsupervised learning paradigm. The color indications and movement indications are built into the positioning system, of which movement indications are used to focus the attention of the system.

A SOM-based skin detector was proposed which uses two SOM's(skin only and skin+non skin). It was trained on a set of 500 manually labeled images; several color spaces like the RGB model, HSI model, and TSL models were tested (Brown et al. 2001). The result of the tests indicates that different color spaces didn't show vivid performance changes on the SOM skin detectors. The advantage of the SOM method over the histogram methods is that it requires considerably fewer resources to implement the run time applications

The nonparametric method is advantageous over others as they have faster training and since we find probabilities of the pixel to be skin or non-skin, it is independent of the shape of the skin distribution.

One major disadvantage can be the storage space required and generalizing the training data. For example, if we take RGB color space and quantize 8 bits for each color then we'll have 8x3 which is 24 bits and thus we'll need an array of 2^{24} elements to store skin probabilities which is a lot! Thus to reduce the memory need a coarser colorspace sampling of 128x128x128, 64x64x64, and 32x32x32 is taken into consideration for possible training sparsity (Jones 1999).

3. Parametric skin distribution modeling The major disadvantage of the nonparametric model is the storage it demands and that their performance directly depends on the representativeness of the training set. The Parametric method was introduced to overcome them.

This method provides a more compact skin model representation and has the ability to generalize and interpolate the training data. The methods that will be discussed under the parametric model are common in the sense that they all operate in the color space chrominance plane ignoring the luminance information. The parametric models provide the ability to interpolate and generalize the deficient training data by approximating the skin color distribution with a single or mixture of Gaussian distributions.

Under parametric skin distribution modeling we have:

Figure 3. Different techniques under parametric skin modeling technique

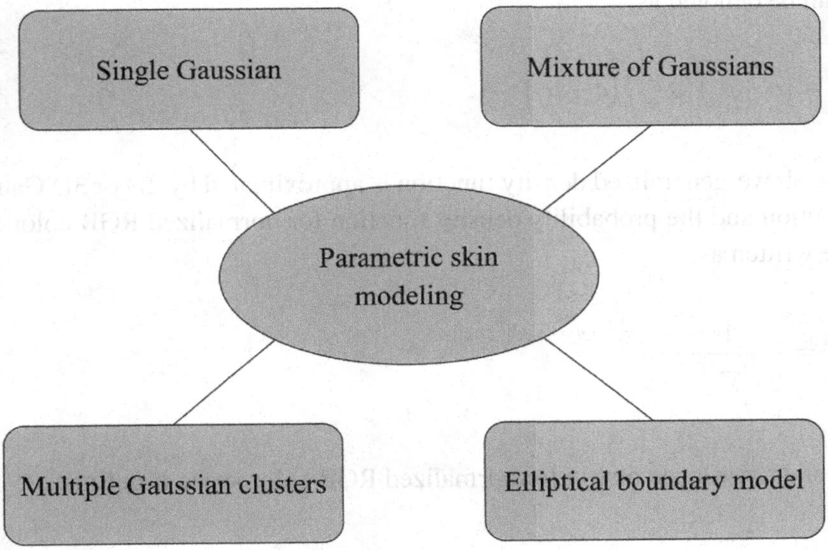

13

a. Single Gaussian method The elliptical gaussian joint probability density function is used to find the skin color distribution. The density function is stated as below:

$$P\left(skin\right) = \frac{1}{2\pi\left|\Sigma_s\right|^{1/2}} \cdot e^{-\frac{1}{2}(c-\mu_s)^T \Sigma_s^{-1}(c-\mu_s)} \quad (17)$$

Where

$$\mu_s = \frac{1}{n}\sum_{j=1}^{n} c_j$$

$$\Sigma_s = \frac{1}{n-1}\sum_{j=1}^{n}\left(c_j - \mu_s\right)\left(c_j - \mu_s\right)^T$$

Here, c is the color vector, μ_s and Σ_s are the mean and covariance matrix respectively, n is the total number of skin color samples c_j. The probability $p(c|skin)$ is the metric to denote how skin-like is the color c.

Mahalanobis distance is used to find the distance between two variables in multivariate space. It even computes the distance between two correlated points for multiple variables. The use of Mahalanobis distance is to find multivariate outliers that can indicate unusual combinations of two or more variables. Thus in skin color detection, the Mahalanobis distance is calculated from color c vector to mean vector given the covariance vector serve the same purpose to denote how skin like is color c, it can be denoted as,

$$\lambda_s\left(c\right) = \left(c - \mu_s\right)^T \Sigma_s^{-1}\left(c - \mu_s\right) \quad (18)$$

The above-generalized density function is approximated by 2D or 3D Gaussian distribution and the probability density function for normalized RGB color space can be written as:

$$\left(skin\right) = \frac{1}{2\pi\left|\Sigma_s\right|^{1/2}} \cdot e^{-\frac{1}{2}\left(\left[r\,r\,g\,g\right]^T \Sigma_s^{-1}\left[r\,r\,g\,g\right]\right)} \quad (19)$$

Mean and covariance matrix for normalized RGB color space is defined as:

$$r = \frac{1}{n} \sum_{j=1}^{n} r_j \tag{20}$$

$$\underline{g} = \frac{1}{n} \sum_{j=1}^{n} g_j \tag{21}$$

$$\Sigma_s = \begin{bmatrix} \sigma_r^2 & \sigma_{rg} \\ \sigma_{rg} & \sigma_g^2 \end{bmatrix} \tag{22}$$

We discussed the general form of Mahalanobis distance above, the formula to calculate Mahalanobis distance for normalized RGB from color point to mean vector is:

$$D = \begin{bmatrix} r\,\underline{r}\,g\,\underline{g} \end{bmatrix}^T \Sigma_s^{-1} \begin{bmatrix} r\,\underline{r}\,g\,\underline{g} \end{bmatrix} \tag{23}$$

Skin segmentation using this method is calculated on the probability achieved based on the probability density function value. This value is determined for each of the pixels.

b.   b. Mixture of Gaussians method The Gaussian mixture model is a more sophisticated model for describing complex-shaped distributions. The probability density function is:

$$p(c \mid skin) = \sum_{i=1}^{k} \pi_i \cdot p_i(c \mid skin) \tag{24}$$

Here k is the count of mixture components, π_i is the mixing parameter and $p_i(c \mid skin)$ is the Gaussian probability density function having its own mean and variance matrices. Skin pixel classification is performed by comparing the $p(c \mid skin)$ to some threshold.

The value of k is a hyperparameter and the choice of k is very crucial as the model needs to avoid overfitting. Expectation maximization techniques are used for model training which assumes the number of components k decided beforehand by Jones (1999) and Yang and Ahuja (1999). In the paper by Yang and Ahuja (1999), the researchers have used the number of components as k=2 whereas in Jones (1999) they have used it as 16. After a lot of research, the bootstrap test k=8 by Terrillon

15

et al. (2000) was chosen as the "good compromise between accuracy of estimation of the true distributions and computational load for thresholding".

This model is computationally complex and thus increases the number of probability density functions considered for modeling.

c. Multiple Gaussian clusters method: This technique is a variant of the k-means clustering algorithm, where the k-means for Gaussian clusters perform model training by approximating skin clusters with 3D Gaussian in YCbCr color space. Mahalanobis distance as discussed earlier is used for classifying the skin pixel as skin or non-skin. Mahalanobis distance is calculated from the color vector to the model cluster center and classification is done using the predefined threshold. This method, unlike other methods, doesn't operate in the color space chrominance plane and does even consider luminance information.

d. d. Elliptic boundary model method: This method has been widely used for research and application purposes. After examining the skin and non-skin models, the researchers Lee and Yoo (2000) came to the conclusion that the skin cluster has an elliptical shape which is not identified by the gaussian models that are discussed in the above points. The Gaussian model does not capture the asymmetry of the skin cluster with respect to its density peak and thus it gives false-positive results and this is how the elliptical boundary model came into use. This model is equally fast and simple and is defined as:

$$\phi(c) = (c - \phi)^T \Lambda^{-1} (c - \phi) \qquad (25)$$

The model training using this method involves two steps. In the first step color images with low frequency are eliminated which accounts for up to 5% of the training data so as to remove the unnecessary noise and negligible data. $\phi \, and \, \Lambda$ is called the model parameters which are calculated as:

$$\phi = \frac{1}{n} \sum_{i=1}^{n} c_i \qquad (26)$$

$$\Lambda = \frac{1}{N} \sum_{i=1}^{n} f_i \cdot (c_i - \mu)(c_i - \mu)^T \qquad (27)$$

Here, n denotes the total number of distinctive training color vectors c_i of the training skin pixel set, f_i is the number of skin samples of the color vector. The N and μ mentioned above while computing the model parameters are calculated as:

$$= \frac{1}{N} \sum_{i=1}^{n} f_i c_i \qquad (28)$$

$$N = \sum_{i=1}^{n} f_i \qquad (29)$$

Again, a threshold value is used to compare the pixels so as to classify them as skin and non-skin pixels. If the pixel color is taken as c then the $\phi(c) < \theta$, where θ is the threshold value. The authors claim that this method calculates better skin clusters as the skewed data does not affect the centroid ϕ calculation.

REAL-LIFE APPLICATIONS OF SKIN COLOUR BASED SEGMENTATION

Skin colour is a feature of the human body which has found many applications in the world of technology. This unique property of colour is used as an advantage in many applications which makes human lives easy as well as automates tasks. Skin colour based segmentation has found various applications in the field of technology out of which many are listed below.

Hand Segmentation

Hand gesture recognition has found a lot of applications in real life, especially in the now-changing world. There are many applications that provide utility to people who are physically challenged too. For a system of this kind to perform efficiently, it becomes necessary to first segment the hand from the live feed of the video which is done using skin colour-based segmentation techniques. This allows the system to map the gesture performed by the hand with the corresponding action.

Following are some of the applications of Hand Gesture Recognition:

1. Media Player Control using Hand Gestures: One of the very newfound applications of hand gesture recognition would be controlling media using hand

gestures. There are various API's that developers use to control the media of the system. Following gestures can be implemented for a media player:

   a.   Swipe right with one finger to fast forward the content.
   b.   Swipe left with one finger to rewind the content.
   c.   Swipe up with two fingers to increase the volume.
   d.   Swipe down with two fingers to decrease the volume.
   e.   Making a fist for muting the volume.
   f.   Opening the first for getting the volume to 100 percent.

2.   Photo Viewer Control using Hand Gestures: Another application for hand gesture recognition would be controlling a photo viewer app using hand gestures. Following gestures can be implemented for a photo viewer application:

   a.   Swiping left to go one photo behind.
   b.   Swiping right to go one photo ahead.
   c.   Making a thumbs-up gesture to mark the photo as a favorite

3.   Hand Gesture Recognition in Online Meetings: One of the recent applications of recognizing hand gestures would be the ability to recognize gestures in online meetings. The recent development of the coronavirus pandemic has given birth to this new feature and many online meeting applications have started using this feature and have incorporated it into their own application.

This basically consists of recognizing the gestures made by the user and converting them into emoji forms so as to give the reaction to the presenter in a subtle format. For example, one makes a thumbs-up sign and the thumbs-up emoji is displayed to everyone. Like this, another example would be one making a clapping hands gesture and that gesture is then recognized and a sound, as well as an emoji associated with it, is played and displayed respectively.

This leads to higher interactivity of the user during the online presentations and gives it a more aligned feel with the offline mode of presenting.

4.   Sign Language Recognition using Hand Gestures: Sign Language Recognition is also one of the applications for hand gesture recognition. This allows mute people to communicate in an easier format with people who do not understand sign language. This also helps people who are trying to learn sign language so they can try out various signs and understand which sign corresponds to which emotion.

5.   Volume modulation using Hand gestures: Over time, analysis in varied fields of computer science has increased the capabilities of sensible homes with refined convolutional machine learning models that constantly analyze sound input for activation phrases and context-dependent correction of detected words and

phrases in commands. Yet the applications based on gesture recognition stay as research work. The hand tracking system has its own set of barriers as:

a. The dimension of the input
b. Clarity of the video input
c. Presence of multiple hands
d. Skin texture and color of the hands
e. Associating functions to unique hand gestures

For the topic of volume control using hand gestures the pointers to the index finger and thumb are used. These tasks are performed using Image and video processing by manipulating the image as input and locating the object of importance. This application presents more viable and easy-to-understand strategies for human-computer connection insightfully with the use of hand motions.

The methodology has a hand tracking module that tracks down the palm and locates points on the palm called the landmarks. These landmarks are represented by dots and joined by the line which measures the length of the difference between the position of the index finger and the thumb to calculate the volume indication as a hand gesture by the user.

In smart home devices, there will be an in-build camera that inputs video as frames per second. The application identifies the hand movements and as soon as the user freezes at a point, that volume is set to the device.

Some of the observations say that the volume modulation works pretty well in general scenarios but some of the problems were noticed as follows:

- The distance of the hand from the camera is inversely proportional to the accuracy of the change in volume i.e. the further away the hand is from the camera, the more difficult it is for the model to predict the change in length accurately.
- The naive implementation was a bit too sensitive so we reduced the sensitivity with some fine-tuning.

The result are shown in Figures 4 and 5.

Figure 4. Output at volume 0

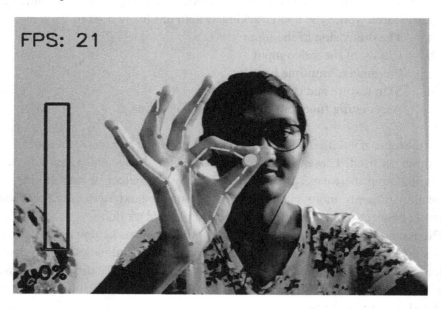

Figure 5. Output at volume 100

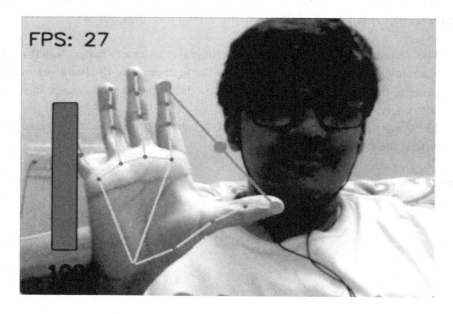

Apart from hand gesture recognition, skin-color based segmentation is also used in other applications like the following:

1. Image Content Filtering: There is increasing access to a lot of uncensored content on the internet for everyone to view. There are many cases where children get exposed to adult images that get circulated via various mediums like chat applications, social media applications, or the entire internet in general. Skin colour-based segmentation helps find out the content in these pictures and can classify them accordingly if they contain any adult content whatsoever. Many papers have also been written for the implementation of such a system. Thus, using these implementations, children can remain safe from the uncensored and explicit content shown on various platforms.

2. Content-Aware Video Compression: In today's fast-changing world, we are constantly exchanging information in the form of video and images. Needless to say, video content occupies and requires a large amount of data for storing and transmitting respectively. Thus, video compression has become a necessity today. As, lower the amount of data, the lower is the cost of storing it and lower is the cost of transmitting it and the higher is the speed and efficiency of transmission. Hence, several methods have been implemented for video compression with content-aware video compression being one of them. In content-aware video compression, what basically happens is that after the conversion of video to frames, the background in each frame is subtracted from the previous frame and a new frame with a blurred background is obtained. So what happens is that the model is aware of the content in the video and segments it into the foreground so the main information is not lost and at the same time the video gets compressed. Skin colour based segmentation helps in segmenting the human objects from the background and thus helps in the crucial process of content-aware video compression.

Figure 6. Flow chart of content-aware video compression

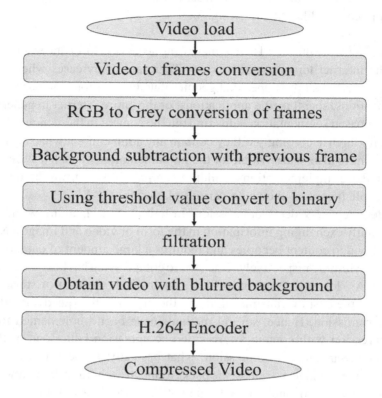

3.   Image Color Balancing: The number of people using a smartphone is increasing exponentially day by day and the demand for professional photo capturing devices is decreasing accordingly.

A high contributing factor to this decline would be that smartphones have come a long way in terms of the quality of photos captured. The photos captured from a camera phone just a mere 5 or 10 years ago significantly differ from the quality of photos captured with phones manufactured now. This increase in quality is the causation of not only the increase in the quality of camera lenses and the decrease in their size but also the ability to post-process the image. This property is used in various photo-editing applications to enhance the quality of the image captured. Using skin colour based segmentation, the highlights of only the portion containing the skin can be refined. This leads to a significant increase in the quality of the overall image.

4.   Face Recognition and Tracking: Humans have been trying to increase the amount of security for safeguarding their personal information. There have been various approaches to it like the fingerprint scanners, pin code protection, etc. In recent years, applications of facial recognition have found significant and tremendous use in various fields. Especially in the ever-changing world after the coronavirus pandemic, there has been a significant rise in the need for facial recognition since the traditional methods like fingerprint scanner requires contact. Skin colour-based segmentation helps in extracting the facial details from an image and excluding other redundant details. This leads to better accuracy in recognizing the face as well as tracking it.

FUTURE RESEARCH DIRECTIONS

As seen in the above discussions, this chapter talked about various aspects of Skin colour-based hand segmentation and various real-life applications of skin colour as a feature for different applications as well as the general applications of hand segmentation. The future of the study of the use of skin-colour is vastly underexplored and there is a lot of area for improvement on the existing techniques as well as inventions of new applications of the same. For example, there is a need for improvement in the scenarios where the background also has a colour similar to the colour of the skin, which leads to unpredictable results. Other areas of improvement would be the speed of the system which uses hand gesture recognition as more delay in the recognition of the gesture leads to an unpleasant experience for the user and eventually leads to the discontinuation of the use of the system. Apart from this, there are many areas where skin colour based segmentation can be improved like techniques of machine learning and deep learning can be added to the processing pipeline which helps in explicitly defining the skin region to use for the further process flow of segmenting the skin and non-skin pixels. This may lead to much better results in terms of both speed and accuracy. Also, various techniques still lack the integration of a dynamic skin-distribution model which is needed as the skin color varies according to different ethnicities. So more research can be done on this aspect too as integrating this dynamic model will lead to much better accuracy and a good generalization of the overall pipeline.

CONCLUSION

In this chapter, the description, comparison and evaluation results of popular methods for skin colour based hand segmentation techniques have been provided.

The most notable and significant differences between the methods, their advantages and disadvantages have been summarized. The most important conclusions drawn from this chapter are listed below:

- Parametric skin modeling methods are suitable for constructing classifiers in case of limited training and expected target data set. The generalization and interpolation ability of these methods make it possible to construct a classifier with an acceptable performance from incomplete training data.
- The methods that are less dependent on the skin cluster shape and take into account skin and non-skin colors overlap (Bayes SPM, Maximum entropy model, automatically constructed colorspace, and classification rules look more promising for constructing skin classifiers for large target datasets.
- Excluding color luminance from the classification process cannot help achieve better discrimination of skin and non-skin colors, but can help to generalize sparse training data.
- Evaluation of colorspace goodness by assessing skin/non-skin overlap, skin cluster shape, etc. regardless of any specific skin modeling method cannot give the impression of how good is the colorspace suited for skin modeling, because different modeling methods react very differently on the colorspace change.

REFERENCES

Albiol, A., Torres, L., & Delp, E. J. (2001). Optimum color spaces for skin detection. *Proceedings - International Conference on Image Processing, 1*, 122–124.

Birchfield, S. (1998). Elliptical head tracking using intensity gradients and color histograms. *Proceedings of CVPR '98*, 232–237. 10.1109/CVPR.1998.698614

Brand, J., & Mason, J. (2000). A comparative assessment of three approaches to pixel evel human skin-detection. *Proc. of the International Conference on Pattern Recognition*, 1, 1056–1059. 10.1109/ICPR.2000.905653

Brown, D., Craw, I., & Lewthwaite, J. (2001). A som based approach to skin detection with application in real time systems. *Proc. of the British Machine Vision Conference*. 10.5244/C.15.51

Chen, Q., Wu, H., & Yachida, M. (1995). Face detection by fuzzy pattern matching. *Proc. of the Fifth International Conference on Computer Vision*, 591–597. 10.1109/ICCV.1995.466885

Devi & Thakur. (2017). Content Aware Video Compression:-An Approach To VOS Algorithm. *International Journal of Trend in Research and Development, 4*(4).

Gomez, G. (2000). On selecting colour components for skin detection. *Proc. of the ICPR, 2,* 961–964.

Gomez, G., & Morales, E. (2002). Automatic feature construction and a simple rule induction algorithm for skin detection. *Proc. of the ICML Workshop on Machine Learning in Computer Vision,* 31–38.

Jones, M. J., & Rehg, J. M. (1999). Statistical color models with application to skin detection. *Proc. of the CVPR '99, 1,* 274–280. 10.1109/CVPR.1999.786951

Lee, J. Y., & Yoo, S. I. (2002). An elliptical boundary model for skin color detection. *Proc. of the 2002 International Conference on Imaging Science, Systems, and Technology.*

Schumeyer, R., & Barner, K. (1998). A color-based classifier for region identification in video. In *Visual Communications and Image Processing 1998* (Vol. 3309, pp. 189–200). SPIE.

Shin, M. C., Chang, K. I., & Tsap, L. V. (2002). Does colorspace transformation make any difference on skin detection? *IEEE Workshop on Applications of Computer Vision.* 10.1109/ACV.2002.1182194

Sigal, L., Sclaroff, S., & Athitsos, V. (2000). Estimation and prediction of evolving color distributions for skin segmentation under varying illumination. *Proc. IEEE Conf. on Computer Vision and Pattern Recognition, 2,* 152–159. 10.1109/CVPR.2000.854764

Soriano, M., Huovinen, S., Martinkauppi, B., & Laaksonen, M. (2000). Skin detection in video under changing illumination conditions. *Proc. 15th International Conference on Pattern Recognition, 1,* 839–842. 10.1109/ICPR.2000.905542

Stern, H., & Efros, B. (2002). Adaptive color space switching for face tracking in multi-colored lighting environments. *Proc. of the International Conference on Automatic Face and Gesture Recognition,* 249–255. 10.1109/AFGR.2002.1004162

Terrillon, J.-C., Shirazi, M. N., Fukamachi, H., & Akamatsu, S. (2000). Comparative performance of different skin chrominance models and chrominance spaces for the automatic detection of human faces in color images. *Proc. of the International Conference on Face and Gesture Recognition,* 54–61. 10.1109/AFGR.2000.840612

Vezhnevets, V., Sazonov, V., & Andreeva, A. (2003). *A Survey on Pixel-Based Skin Color Detection Techniques*. In International Conference Graphicon 2003, Moscow, Russia.

Yang, M., & Ahuja, N. (1999). Gaussian mixture model for human skin color and its application in image and video databases. *Proc. of the SPIE: Conf. on Storage and Retrieval for Image and Video Databases (SPIE 99)*, 3656, 458–466 10.1117/12.333865

Zarit, B. D., Super, B. J., & Quek, F. K. H. (1999). Comparison of five color models in skin pixel classification. *ICCV'99 Int'l Workshop on recognition, analysis and tracking of faces and gestures in Real-Time systems*, 58–63. 10.1109/RATFG.1999.799224

Chapter 2
Hand Gesture Recognition Through Depth Sensors

Bhupesh Kumar Dewangan

ⓘ https://orcid.org/0000-0001-8116-7563

School of Engineering, Department of Computer Science and Engineering, O.P. Jindal University, Raigarh, India

Princy Mishra

Rungta College of Engineering and Technology, India

ABSTRACT

A hand gesture recognition system allows for nonverbal communication that is natural, inventive, and modern. It can be used in a variety of applications involving human-computer interaction. Sign language is a type of communication in which people communicate using their gestures. The identification of hand gestures has become one of the most important aspects of human-computer interaction (HCI). The goal of HCI is to get to a point where human-computer interactions are as natural as human-human interactions, and adding gestures into HCI is an essential research topic in that direction. The history of hand gesture recognition is initially presented in this chapter, followed by a list of technical challenges. Researchers from all over the world are working on developing a reliable and efficient gesture recognition system, particularly a hand gesture recognition system, for a variety of applications. In this chapter, the core and advanced application areas and various applications that use hand gestures for efficient interaction have been addressed.

DOI: 10.4018/978-1-7998-9434-6.ch002

INTRODUCTION

The gesture is a type of nonverbal communication, and it is the oldest form of communication between humans. The gesture involves the use of several body parts, primarily the hand and face. Gestures were used by ancient men to express information such as food/prey for hunting, supply of water, knowledge about their enemy, a plea for help, and so on. Gestures are bodily actions that communicate information. If we will not consider the world of computer study about the contact between humans for a while, we may readily see that our communication uses a wide range of gestures. Hand gesture has become a fantastic option for expressing simple concepts, interpreted by the system of gesture recognition, and transformed into events. The possibility of interacting with computer systems to provide meaningful information for the interaction with human-computer is presented in various hand movements classified on the basis of the distinctive shapes of hand, finger, or orientation. Advanced technologies require that varied hand gestures and their recognition, classification, and interpretation so they can be applied in a wide range of computing applications.

The process of automatically recognizing and analyzing human behaviors from data is known as human action recognition, which can be acquired from various types of sensors such as RGB cameras, depth cameras, range sensors, wearable inertial sensors, or other modality-type sensors. Nowadays gestures are widely used for many applications in various fields. As a novel input modality in human-computer interaction, gesture recognition systems enable a more intuitive and convenient means of interacting. In human-computer interaction, hand gesture recognition provides the fewest limitations for the user (HCI). Hand gestures have a wide variety of applications in Human-Computer Interaction (HCI), which can ensure speed of communication and provide an easier user-friendly and aesthetic environment, provide non-professional contact with the computer from a distance for comfort and security, and control complex and virtual environments in a much easier approach (Andrea, 2014). In contrast, the application of hand gestures requires the user to be an expert and well-trained to utilize and grasp the meaning of various gestures (Samata & Kinage, 2015). A group of manual gestures can be employed for each application to complete its functions. From the point of the computer of view, the external factors, such as light, skin color, location, and orientation of the hand affect the performance of the recognition of hand motions (Vaibhavi et al., 2014). A classification of gesture recognition techniques has been shown in figure 1.

Figure 1. Classification of gesture recognition

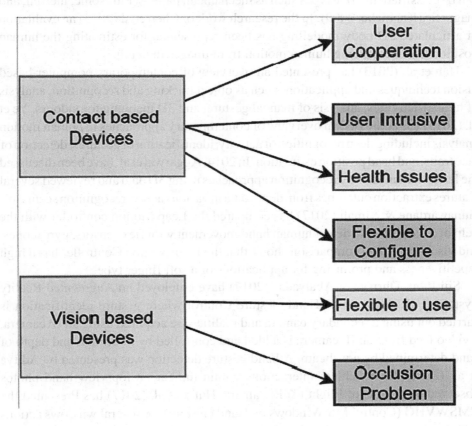

BACKGROUND

In the research field, hand gesture recognition has been introduced for 38 years (Prashan, 2014). A hand glove was proposed by Zimmerman in 1977 that was able to detect the number of fingers bending (Praveen & Shreya, 2015). In 1983 Gary Grimes has developed a system to determine if a thumb or finger touches any part of the hand (Laura, Angelo & Paolo, 2008). The word gesture recognition refers collectively to an entire process of recording human motions into semantic-significant orders for their representation and conversion. Human-computer interaction is dependent on two basic technologies which are named contact-based and vision-based. The physical user interaction with the interfacing device is a foundation of contact-based devices used for hand gesture detection systems. The interfacing devices can be assembled with various devices such as multi-touch screens, data gloves, and accelerometers, etc. Contact-based devices for hand gesture recognition can be classified into five types such as mechanical, haptics, ultrasonic, inertial, and magnetic (Kanniche, 2009). In the research work of Chen et.al (2013) an evaluation of articulated 3D body modelling has been represented for estimating the human position and recognizing human motion from images in depth.

Han et al. (2013) has presented an overview of contemporary computer-based vision techniques and applications such as object tracking and recognition, analysis of human activities, analysis of manual gesture, and 3D mapping for indoors. Ye et al. (2013) has addressed an overview of contemporary approaches to human motion analysis including deep recognition of activity, identification of gestures, detection of face traits, and head position estimation. In 2014, Aggarwal et al. have been discussed the latest human activity recognition approaches using 3D data and reviewed several features extraction outcomes from depth data in various action recognition scenarios. Gunawardane & Nimali (2017) has compared the Leap Motion controller with the help of a data glove to detect human hand movement with flex sensors, gyroscopes, and visual data. The comparison shows that the Leap Motion Controller has a high repetitiveness and promising for applications of a soft finger type.

Siji Rani, Dhrisya & Ahalyadas (2017) have employed an Augmented Reality System (HGCARS) novel Hand Gesture Control where gesture identification is carried out using a secondary camera and realities are acquired using an IP camera, a video feed from an IP camera is added and controlled by a location and depth of hand determined by a webcam. A hand gesture detection was presented by Aditya et al. (2017), for touchless interactions within the car on top-view hand photos observed by a Time of Flight (ToF) camera. Haf z et al. (2017) has Presented the CMSWVHG (Control MS Windows by Hand Gesture) for several windows actions utilizing manual gestures for input with the aid of OpenCV using an internal or external camera. (Rishabh et al., 2016) states that One camera source was required to control the OS on the projected screen for a virtual mouse system without the need for hardware.

A real-time EMG-based system was proposed by Marco et al. (2017a), Marco et al. (2017b), Karthik et al. (2017), and Stefano et al. (2018) with the commercial Myo sensor monitored surface electromyography, which is a bracelet implanted in the forearm. In the gesture recognition system, Yifan et al. (2018) have proposed the use of wearable devices such as VR/AR cask and glasses. Jiajun Z, Zhiguo S. (2017) states that for a large library of hand motions signals, a Doppler-radar sensor with double receiving channels can be used. In the work of Yuntao et al. (2017) and Piotr, Tomasz & Jakub (2017) A mechanomyography-based interface (MMG) was used to collect arm and hand gestures. Ibrahim et al. (2018) have Suggested a hand gesture system to measure the symbols at different locations with two monopole antennas. Zengshan et al. (2018) has Developed a Wasatch wireless sensor, which develops the movement locus of the action, building a virtual antenna array based on signal samples.

(Yangke Li, et. Al.,2021) have proposed a novel framework for skeleton- based dynamic hand gesture recognition. They have designed a compact joint encoding method in the spatial perception stream (SP-Stream). This system can adaptively select compact joints based on the convex hull of the hand skeleton and encode them into a skeleton image for fully extracting spatial features. Moreover, they have also presented a global enhancement module (GEM) to enhance key feature maps. Authors have also proposed a motion perception module (MPM) in the temporal perception stream (TP-Stream), to enhance the notable movement of hand gestures on coordinate axes.

Hand gesture recognition can be done by acquisition methods, feature extraction, and classification techniques. Where the acquisition method uses surface electromyography sensors with wearable devices, feature extraction can be done by using histograms and classification can be performed by using an artificial neural network.

The input for hand gesture identification was collected by vision-based devices in human-computer interaction. The system uses one or more cameras to interpret and analyze the movements using a video sequence (Mitra and Acharya 2007). Vision-based hand gesture recognition technique also uses hand markers for the detection of human hand motion and gestures. The hand markers can be divided into two categories such as reflective markers and LED lights, where reflective markers are passive, and flashes as strobes where LED lights are active and flash in sequence. The major issue of vision-based hand recognition is to face a wide range of actions and gestures. It must also balance the compromise of precision-performance-usefulness by application type, the cost of the solution, and a number of factors such as performance in real-time, resilience, scalability, etc. Since the contact-based system can be uncomfortable for users since they involve physical contact with the user, it is still possible for these devices to be precise to recognize and less complex to deploy. Health concerns are the main downside of contact-based devices. Vision-based appliances are however user-friendly but have problems with settings and occlusion are caused by their supporting devices such as mechanical sensor materials that elevate allergy symptoms, magnet devices that increase the cancer risk, etc (Schultz et al. 2003), (Nishikawa et al. 2003). Some criteria-based comparisons of contact-based and vision-based hand gesture recognition has shown below figure.

Figure 2. Comparison between contact and vision-based devices

Vision-Based Hand Gesture Recognition

Vision-based hand gesture recognition system got more attention from researchers because it has many advantages such as a human hand can work as an input for the system and no intermediate devices are required. So, the user would be able to control the system with their gestures which are already defined to the system previously. Gestures image sequences can be captured with the help of a conventional RGB camera by processing and analyzing the images.

Chu et. Al. (2011) have presented a vision-based hand motion gesture with a self-portrait interface. Their research enables a user in self-portrait photos to manipulate a digital camera. a vision-based sign language translation device has been developed by Madhuri y. et al (2013) to help people with hearing or speech impairments to communicate with listeners. It's employed as a translator for people who don't understand sign language, avoids intermediate interference, and allows communication using their natural style of speech. The real-time performance is good as compared with other devices. This system makes it possible the instant recognition from finger and hand motions to translation and users could be able to recognize the alphabet (A-Z) and the number of the handed signs (0-9) as mentioned in figure 2 and figure 3.

Figure 3. Database for alphabet signs
(Chen, L.,2013)

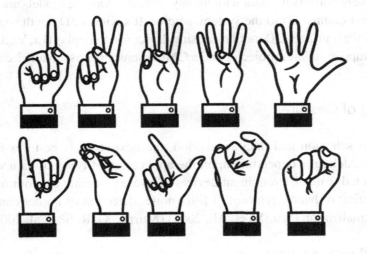

Figure 4. Database for number signs
(Chen, L.,2013)

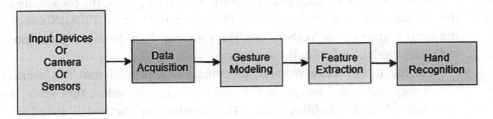

Basic Steps of Hand Gesture Recognition

Input Acquisition

Retrieving input data with the help of input devices. A multitude of data collecting input devices is available. Some are data guides, markers, drawings, and pictures taken from the stereo/webcam/Kinect 3D sensor. Wireless data handles are currently commercially available so that the obstruction by the cord can be removed. Color markers are connected to the skin of the human being, which are also utilized for taking input and color localization by hand. Input can also be provided to the system, except for a low-cost Internet camera, without any costly additional hardware. The single or double hand is utilized for generating hand gestures, and the camera readily and intuitively collects the data without any contact. Drawing models are also used to give input commands to the system. Microsoft's Kinect 3D depth sensor is the most commonly utilized 3D motion-sensing device for gameplay (Li, Y.,2019). This system comprises a laser projector and a CMOS sensor for use in all circumstances of lighting.

Modeling of Gestures

After input selection and data acquisition, the next stage is gestures modeling. Gestures modeling is depending upon the application of types of data which has been collected from different input devices. Gesture modeling has four basic steps as Segmenting of hands, removal of filter/noise, detection of border/contour, and finally normalization (Wu, Y., et. Al., 2001) (Murthy, G. R. S., et al, 2009).

- **Hand segmentation:** Hand segmentation is also known as localization of hand(s), refers to the location of hands or hand sequence in the images. Hand segmentation methods classify the image into foreground and background homogeneous parts, where foreground contains the object which is hand the background contains the rest part of the image. When acquiring data utilizing contact-based devices, such as keyboard-mouse, data gloves, the background doesn't matter (Lamar, M. V., et al, 1999) (Liang, R. H.,et. Al,1998). Some of the hand segmentation methods are Thresholding, Skin-based, Subtraction, statistical model, and color normalization.
  - In the thresholding method, the image is separated into two areas, background, and foreground, based on a specific value, this value is termed as thresholding value. The thresholding method is based on many real-time characteristics, some of which utilize a range/profundity threshold, while some use a color threshold (RGB or HSV). Otsu's thresholding (Guo, J., & Li, S., 2011) is another essential way to histogram-based picture thresholds or to reduce the grey image level to a binary image provided that only two types of pixels exist in the image.
  - Human skin colors, which are also known as hue, saturation, and brightness and it is abbreviated as HSB, can be used to separate the hand, head, or body from the background with an RGB or grey, or HSV color (HSB) representation. This technique could be used to differentiate the object and background from the previously selected color range.
  - To discover any changes in the test image, the backdrop of a test image is separable by reference to an earlier static image, and this segmentation method is known as subtraction. This method shows the best results in

the extraction of moving objects in videos captured with a static camera. A robust approach of background subtraction should be able to handle lighting fluctuations, repeating movements from clutter, and changes to long-term scenes.

○ The image is turned into a statistical model to assign a probabilistic loyalty to each pixel. For achieving statistical model Bayesian rule (Bhuyan, M. K.,2012), Gaussian mixture model (Hasan, M. M., & Mishra, P. K. 2012), and Expectation-Maximization (EM) (Jeong, M. H.,2002) algorithms are in use.

○ Illumination factors can be varied according to lighting factors and camera also, so the Color normalization technique makes it possible to adjust for these differences for object identification systems based on RGB color (Lamar, M. V.,1999).

- **Noise Reduction:** Noise reduction is an important part of hand gesture recognition because, during image capture processing or the transmission of the image, noise can be added to the original image. Noise can be affecting the quality of the image, so noise reduction can be done with the help of various filters, some of which are salt and paper filters, morphological operation, and multidimensional mean.

  ○ When white and black pixels randomly appear in any image salt and a paper filter are used for the reduction.

  ○ The morphological operation can also be used to reduce noise in a better way. Morphological operation is a mathematical operation if set theory and can be done into two parts first is erosion and dilation. An erosion filter (Lamar, M. V., et al,1999) (Meena, S., 2011) (Elmezain, M.,et. al., 2007) tends to lower the size of bright pixels by correlating them with neighboring dark areas, but the dilation filter does the reverse and therefore reduces the dark pixel.

  ○ A collection of eight filters is utilized here to detect the edge noise or to fluctuate for greater edge smoothness instead of a single filter.

- **Contour Detection:** Contour detection is also known as edge detection of the image, which helps to capture the significant properties of any image object. According to Maini, R., & Aggarwal, H. (2009) there are several edge detection methods are available, they might be classed as edge detection based on gradients and edge detection based on Laplacian. The method which is based on gradient helps to find the borders in the first derivative of the image, which determines the zero crossings in the second derivative of the image to detect the edges.

- **Normalization:** The penultimate phase in gesture modeling is normalization or reduction of space. Since the area of interest for any image is confined in a small region, just the corresponding area should be crop and then further processed. Some of the normalization techniques are cropping operation, dimension unification, and significant feature location.

- ○ In cropping the appropriate and meaningful area of the image would be cropped and fitted into a picture window without unnecessary background (Panwar, M. (2012) (Kumarage, D.,2011) (Fang, Y.,2007) This technique is used to make the image more meaningful and relevant.
- ○ Dimension unification is a technique to set the image size uniformly to a specific dimension. This dimension could be varied according to the application, research, and type of hardware, etc.
- ○ Unnecessary features of an image should also be reduced because they can affect computation time. For the extraction of features, some researchers have used several machine learning techniques. For example, Stergiopoulou, E., & Papamarkos, N. (2009) have used the neural network for reducing function space.

Feature Extraction

For hand gesture recognition, a wide range of parameters may be used including shape, orientation, texture, shape, outline, motion, distance, the center of gravity, etc. and all the features can play a very important role in hand gesture recognition. Human-driven gestures, such as hand contour, fingertips, and finger detections, may be identified by geometrical characteristics. But because of occlusions and illuminations (Murthy, G. R. S., & Jadon, R. S. 2009), these features cannot always be available or dependable.

There can also be recognition of some non-geometric functions such as color, silhouette, the texture is possible, but for the purpose they are insufficient. To choose the features automatically and implicitly, the processed picture or image may be provided to the recognizer instead of employing the single feature by itself (Murthy, G. R. S., & Jadon, R. S. 2009). Mainly three approaches could be used for features extraction first one is the Model-based Kinematic model (Rajesh, R. J., et al, 2012) approach second is the view-based approach and the third is the low-level feature-based approach (Lahamy, H., & Lichti, D. 2012).

Hand Gesture Recognition

After extracting the relevant features from the photos, and selecting the right data set, the gestures can be identified by a standard machine learning or a specially applied classifier. From the researchers' point of view some hand gesture recognition methods that are widely in use are as follows:

Hidden Markov (HMM) model (Yang, Z., et al,2012) (Oka, K., et al,2002) (Kim, J. B., et al,2002) (Yoon, H. S., et al,2001) (Min, B. W., et al, 1997) for dynamic hand gesture recognition data containing temporal information. The best property of HMM model is the high recognition rate. Kasprzak, W., et al, (2012) have stated that HMM can also be implemented with other classifiers (Elmezain, M., et al,2010). Another most famous technique is the support vector machine also known as SVM. SVM is based on supervised learning, so it can take input data and predict the output. Some researchers have used it without classifiers such as (Sultana, A., & Rajapuspha, T. 2012) (Geetha, M., & Manjusha, U. C. 2012) (Pradhan, A., et al, 2012), some has used it in the form of multiclass SVM (Dardas, N. H., & Georganas, N. D. 2011) (Rekha, J., et al, 2011)(Dardas, N., et al, 2010), Some of them are using the infused form (Rekha, J., et al, 2011)(Meena, S. 2011).

Depth Sensors

The Depth sensors can be able to track monocular clues as well as actual depth measurements of the environment which can be called depth clues. Hand Gesture recognition is one of the best applications of depth sensors. The data of a deep sensor/ camera offers information on the curvature of your hand, the shape of the hand from the centre of the palm, and the size and number of hand components in your gesture throughout each point on your hand contortion. According to Domino F. et al, (2014), a combination of leap motion and depth sensor with a gesture recognition system allows hand movements to be identified. The depth images generated from depth cameras show insensitivity to change in light and have resulted in a good performance in the identification of human action. To increase the performance of human gesture recognition systems for combining data from these two modal sensors because of the additional information provided by depth images of a depth sensor and by inert signals from a wearable sensor. The depth images collect movement properties globally whereas inertial signals capture the attributes of the local movement (Chen, C.et. Al,2017).

Advantages of Depth Sensors

- Depth sensors are economic and widespread available.
- Depth sensors are Insensitive to lights and brightness fluctuations and can easily work in the dark.
- It may provide information on the scenario in the 3D framework.
- Depth sensors are easily operable.
- Depth sensors can work perform properly in any alteration of color and texture because it is non-sensitive towards color and texture modification.

Disadvantages of Depth Sensors

- The basic requirement of depth sensors is that the object should be present on the view plane.
- There is a possibility of noise occurring in the captured image.
- Depth information is sensitive to reflective materials which can absorb light or can be transparent also.
- Depth sensors don't produce any color-based information.

In recent years, various methods for the recognition of human behaviour from profound images have been proposed. Deep sequences of human actions are a bag of 3D points (Li, W., et. Al.,2010), projected depth maps (Chen, C., et.al.,2015), (Chen, C., et.al.,2016) (Yang, X., et al,2012) occupancy patterns (Vieira, A. W.,et. Al,2012), (Wang, J.,et. Al,2012) space-temporal cuboid (Xia, L., & Aggarwal, J. K. 2013), surface normal (Oreifej, O., & Liu, Z.,2013) and skeleton joints (Evangelidis, G.,et. Al.,2014) as the main feature technique for human action detection produced.

Depth Sensors for Hand Gesture Recognition

Following characteristics has been observed while using depth sensors for hand gesture recognition:

- Illumination-invariant for indoor environments, if the environment does not contain light of the same wavelength used for pulse of the sensor.
- Error in depth is approximately 5-15mm for 0.5-3m depth window. Depth information is recorded in the separated depth image (8 bits a pixel). This gives sufficient separation between the hand and the body used for gesture recognition.
- Both depth and color images are captured at real-time speed (30 frames/sec).
- Depth for objects with no texture (such as walls) can still be obtained. However, for certain objects that does not return any signal (e.g., light absorbing materials), distance cannot be obtained.
- Compared to the traditional RGB camera, 3D depth map has significant advantages for its availability to discover strong clues in boundaries and 3D spatial layout even in cluttered background and weak illumination.

FUTURE RESEARCH DIRECTIONS

The usage of hand gesture recognition for communication between the user and the equipment should be expected in applications like domestic control systems, healthcare systems, gaming technologies, Television, home automation, and robotics (Hexa, 2017). Smart televisions are also expected to experience growth in this topic and increase the purchasing rate of the latest technology using hand gestures (Hexa, 2017). The topic is expected to grow over 28% from the year 2017 to 2024 (Hexa, 2017).

CONCLUSION

This chapter successfully covers the most crucial approaches, applications, and various techniques for the recognition of hand gestures. In a way to make this more impactful, various research papers have been reviewed. All the basic steps for hand gesture recognition such as image acquisition, gesture modeling, feature extraction, and last but not the least hand gesture recognition. The result of this chapter can be summarized as, after covering the main advantages and disadvantages of depth sensors, it can be used with the fusion of inertial sensors. According to some comparative study the RGB-based recognition showed an accuracy of 99.54%, and the depth-based recognition showed an accuracy of 99.07%, this suggests that depth-based recognition may be good enough in hand gesture recognition. As a general remark, the simultaneous employment of these two different modality sensors provides for increased identification of human gestures as compared with scenarios in which each sensor is employed separately. Both of the sensors are cost-effective and feasible with 3D data. However, there are still some issues when fusing depth and inertial to recognize measurements.

ACKNOWLEDGMENT

This research received no specific grant from any funding agency in the public, commercial, or not-for-profit sectors.

REFERENCES

Aggarwal, J. K., & Xia, L. (2014). Human activity recognition from 3d data: A review. *Pattern Recognition Letters*, *48*, 70–80. doi:10.1016/j.patrec.2014.04.011

Attwenger, A. (2017). *Advantages and Drawbacks of Gesture-Based Interaction.* Grin Verlag.

Bhuyan, M. K., Neog, D. R., & Kar, M. K. (2012). Fingertip detection for hand pose recognition. *International Journal on Computer Science and Engineering, 4*(3), 501.

Chen, C., Jafari, R., & Kehtarnavaz, N. (2015, January). Action recognition from depth sequences using depth motion maps-based local binary patterns. In *2015 IEEE Winter Conference on Applications of Computer Vision* (pp. 1092-1099). IEEE. 10.1109/WACV.2015.150

Chen, C., Jafari, R., & Kehtarnavaz, N. (2017). A survey of depth and inertial sensor fusion for human action recognition. *Multimedia Tools and Applications, 76*(3), 4405–4425. doi:10.100711042-015-3177-1

Chen, C., Liu, K., & Kehtarnavaz, N. (2016). Real-time human action recognition based on depth motion maps. *Journal of Real-Time Image Processing, 12*(1), 155–163. doi:10.100711554-013-0370-1

Chen, L., Wang, F., Deng, H., & Ji, K. (2013, December). A survey on hand gesture recognition. In *2013 International conference on computer sciences and applications* (pp. 313-316). IEEE. 10.1109/CSA.2013.79

Chen, L., Wei, H., & Ferryman, J. (2013). A survey of human motion analysis using depth imagery. *Pattern Recognition Letters, 34*(15), 1995–2006. doi:10.1016/j.patrec.2013.02.006

Chu, S., & Tanaka, J. (2011, July). Hand gesture for taking self-portrait. In *International Conference on Human-Computer Interaction* (pp. 238-247). Springer.

Dardas, N., Chen, Q., Georganas, N. D., & Petriu, E. M. (2010, October). Hand gesture recognition using bag-of-features and multi-class support vector machine. In *2010 IEEE International Symposium on Haptic Audio Visual Environments and Games* (pp. 1-5). IEEE. 10.1109/HAVE.2010.5623982

Dardas, N. H., & Georganas, N. D. (2011). Real-time hand gesture detection and recognition using bag-of-features and support vector machine techniques. *IEEE Transactions on Instrumentation and Measurement, 60*(11), 3592–3607. doi:10.1109/TIM.2011.2161140

Dipietro, L., Sabatini, A. M., & Dario, P. (2008). A survey of glove-based systems and their applications. *IEEE Transactions on Systems, Man, and Cybernetics, Part C (Applications and Reviews), 38*(4), 461-482.

Dominio, F., Donadeo, M., & Zanuttigh, P. (2014). Combining multiple depth-based descriptors for hand gesture recognition. *Pattern Recognition Letters*, *50*, 101–111. doi:10.1016/j.patrec.2013.10.010

Elmezain, M., Al-Hamadi, A., Krell, G., El-Etriby, S., & Michaelis, B. (2007, December). *Gesture recognition for alphabets from hand motion trajectory using hidden Markov models. In 2007 IEEE International Symposium on Signal Processing and Information Technology*. IEEE.

Elmezain, M., Al-Hamadi, A., Sadek, S., & Michaelis, B. (2010, December). Robust methods for hand gesture spotting and recognition using hidden markov models and conditional random fields. In *The 10th IEEE International Symposium on Signal Processing and Information Technology* (pp. 131-136). IEEE. 10.1109/ISSPIT.2010.5711749

Erol, A., Bebis, G., Nicolescu, M., Boyle, R. D., & Twombly, X. (2007). Vision-based hand pose estimation: A review. *Computer Vision and Image Understanding*, *108*(1-2), 52–73. doi:10.1016/j.cviu.2006.10.012

Evangelidis, G., Singh, G., & Horaud, R. (2014, August). Skeletal quads: Human action recognition using joint quadruples. In *2014 22nd International Conference on Pattern Recognition* (pp. 4513-4518). IEEE.

Fang, Y., Cheng, J., Wang, K., & Lu, H. (2007, August). Hand gesture recognition using fast multi-scale analysis. In *Fourth international conference on image and graphics (ICIG 2007)* (pp. 694-698). IEEE. 10.1109/ICIG.2007.52

Geetha, M., & Manjusha, U. C. (2012). A vision based recognition of indian sign language alphabets and numerals using b-spline approximation. *International Journal on Computer Science and Engineering*, *4*(3), 406.

Gunawardane, P. D. S. H., & Medagedara, N. T. (2017, October). Comparison of hand gesture inputs of leap motion controller & data glove into a soft finger. In *2017 IEEE International Symposium on Robotics and Intelligent Sensors (IRIS)* (pp. 62-68). IEEE. 10.1109/IRIS.2017.8250099

Guo, J., & Li, S. (2011). *Hand gesture recognition and interaction with 3D stereo camera*. The Project Report of Australian National University.

Hafiz, M. A.-R., Lehmia, K., Danish, M., & Noman, M. (2017). CMSWVHG-control MS Windows via hand gesture. In *International Multi-Topic Conference INMIC*, 1-7.

Han, J., Shao, L., Xu, D., & Shotton, J. (2013). Enhanced computer vision with microsoft kinect sensor: A review. *IEEE Transactions on Cybernetics*, *43*(5), 1318–1334. doi:10.1109/TCYB.2013.2265378 PMID:23807480

Hasan, M. M., & Mishra, P. K. (2012). Hand gesture modeling and recognition using geometric features: a review. *Canadian Journal on Image Processing and Computer Vision, 3*(1), 12-26.

Hexa, R. (2017). Gesture recognition market analysis by technology 2D. 3D. By application tablets & notebooks. Smartphones. Gaming consoles. Smart televisions. In Laptops & desktops and segment forecasts 2014-2024. IEEE.

Ibrahim, A., Hashim, A., Faisal, K., & Youngwook, K. (2018). Hand gesture recognition using input impedance variation of two antennas with transfer learning. *Sensors Journal, 18*, 4129-4135.

Jeong, M. H., Kuno, Y., Shimada, N., & Shirai, Y. (2002, August). Two-hand gesture recognition using coupled switching linear model. In *Object recognition supported by user interaction for service robots* (Vol. 3, pp. 529–532). IEEE.

Kaâniche, M. (2009). *Gesture recognition from video sequences* (Doctoral dissertation). Université Nice Sophia Antipolis.

Karthik, S. K., Akash, S., Srinath, R., & Shitij, K. (2017). Recognition of human arm gestures using Myo armband for the game of hand cricket. *International symposium on robotics and intelligent sensors IRIS*, 389-394.

Kasprzak, W., Wilkowski, A., & Czapnik, K. (2012). Hand gesture recognition based on free-form contours and probabilistic inference. *International Journal of Applied Mathematics and Computer Science*, *22*, 437–448.

Kim, J. B., Park, K. H., Bang, W. C., & Bien, Z. Z. (2002, May). Continuous gesture recognition system for Korean sign language based on fuzzy logic and hidden Markov model. In *2002 IEEE World Congress on Computational Intelligence. 2002 IEEE International Conference on Fuzzy Systems. FUZZ-IEEE'02. Proceedings (Cat. No. 02CH37291)* (Vol. 2, pp. 1574-1579). IEEE.

Kumarage, D., Fernando, S., Fernando, P., Madushanka, D., & Samarasinghe, R. (2011, August). Real-time sign language gesture recognition using still-image comparison & motion recognition. In *2011 6th International Conference on Industrial and Information Systems* (pp. 169-174). IEEE.

Lahamy, H., & Lichti, D. (2012). Robust Real-Time and Rotation-Invariant American Sign Language Alphabet Recognition Using Range Camera. *Proceedings of the International Archives of the Photogrammetry, Remote Sensing and Spatial Information Sciences*, 217-222.

Lamar, M. V., Bhuiyan, M. S., & Iwata, A. (1999, October). Hand gesture recognition using morphological principal component analysis and an improved CombNET-II. In *IEEE SMC'99 Conference Proceedings. 1999 IEEE International Conference on Systems, Man, and Cybernetics (Cat. No. 99CH37028)* (Vol. 4, pp. 57-62). IEEE.

Li, W., Zhang, Z., & Liu, Z. (2010, June). Action recognition based on a bag of 3d points. In *2010 IEEE Computer Society Conference on Computer Vision and Pattern Recognition-Workshops* (pp. 9-14). IEEE.

Li, Y. (2012, June). Hand gesture recognition using Kinect. In *2012 IEEE International Conference on Computer Science and Automation Engineering* (pp. 196-199). IEEE.

Li, Y., Ma, D., Yu, Y., Wei, G., & Zhou, Y. (2021). Compact joints encoding for skeleton-based dynamic hand gesture recognition. *Computers & Graphics*, *97*, 191–199.

Liang, R. H., & Ouhyoung, M. (1998, April). A real-time continuous gesture recognition system for sign language. In *Proceedings third IEEE international conference on automatic face and gesture recognition* (pp. 558-567). IEEE.

Madhuri, Y., Anitha, G., & Anburajan, M. (2013, February). Vision-based sign language translation device. In *2013 International Conference on Information Communication and Embedded Systems (ICICES)* (pp. 565-568). IEEE.

Maini, R., & Aggarwal, H. (2009). Study and comparison of various image edge detection techniques. *International Journal of Image Processing (IJIP)*, *3*(1), 1-11.

Marco, E. B., Andrés, G. J., Jonathan, A. Z., Andrés, P., & Víctor, H. A. (2017a). Hand gesture recognition using machine learning and the Myo armband. In *European Signal Processing Conference*. IEEE.

Marco, E. B., Cristhian, M., Jonathan, A. Z., Andrés, G. J., Carlos, E. A., Patricio, Z., Marco, S., Freddy, B. P., & María, P. (2017b). Real-time hand gesture recognition using the Myo armband and muscle activity detection. In *Second Ecuador technical chapters meeting ETCM*. IEEE.

Meena, S. (2011). *A study on hand gesture recognition technique* (Doctoral dissertation).

Min, B. W., Yoon, H. S., Soh, J., Yang, Y. M., & Ejima, T. (1997, October). Hand gesture recognition using hidden Markov models. In *1997 IEEE International Conference on Systems, Man, and Cybernetics. Computational Cybernetics and Simulation* (Vol. 5, pp. 4232-4235). IEEE.

Mitra, S., & Acharya, T. (2007). *Gesture recognition: a survey. IEEE Trans Syst Man Cybern (SMC) Part C Apple.*

Murthy, G. R. S., & Jadon, R. S. (2009). A review of vision based hand gestures recognition. *International Journal of Information Technology and Knowledge Management, 2*(2), 405–410.

Nishikawa, A., Hosoi, T., Koara, K., Negoro, D., Hikita, A., Asano, S., Kakutani, H., Miyazaki, F., Sekimoto, M., Yasui, M., Miyake, Y., Takiguchi, S., & Monden, M. (2003). FAce MOUSe: A novel human-machine interface for controlling the position of a laparoscope. *IEEE Transactions on Robotics and Automation, 19*(5), 825–841.

Oka, K., Sato, Y., & Koike, H. (2002, May). Real-time tracking of multiple fingertips and gesture recognition for augmented desk interface systems. In *Proceedings of Fifth IEEE International Conference on Automatic Face Gesture Recognition* (pp. 429-434). IEEE.

Oreifej, O., & Liu, Z. (2013). Hon4d: Histogram of oriented 4d normals for activity recognition from depth sequences. In *Proceedings of the IEEE conference on computer vision and pattern recognition* (pp. 716-723). IEEE.

Panwar, M. (2012, February). Hand gesture recognition based on shape parameters. In *2012 International Conference on Computing, Communication and Applications* (pp. 1-6). IEEE.

Piotr, K., Tomasz, M., & Jakub, T. (2017). Towards sensor position-invariant hand gesture recognition using a mechanomyographic interface. In Signal processing: algorithms, architectures, arrangements and applications SPA. IEEE.

Pradhan, A., Ghose, M. K., Pradhan, M., Qazi, S., & Moors, T., El-Arab, I. M. E., ... Memon, A. (2012). A hand gesture recognition using feature extraction. *Int J Curr Eng Technol, 2*(4), 323–327.

Prashan, P. (2014). Historical development of hand gesture recognition. In Cognitive science and technology book series CSAT. Singapore: Springer.

Praveen, K.S., & Shreya, S. (2015). Evolution of hand gesture recognition: A review. *International Journal of Engineering and Computer Science, 4*, 9962-9965.

Rajesh, R. J., Nagarjunan, D., Arunachalam, R. M., & Aarthi, R. (2012). Distance Transform Based Hand Gestures Recognition for PowerPoint Presentation Navigation. *Advances in Computers*, *3*(3), 41.

Rekha, J., Bhattacharya, J., & Majumder, S. (2011). Hand gesture recognition for sign language: A new hybrid approach. In *Proceedings of the International Conference on Image Processing, Computer Vision, and Pattern Recognition (IPCV)* (p. 1). The Steering Committee of The World Congress in Computer Science, Computer Engineering and Applied Computing (WorldComp).

Rekha, J., Bhattacharya, J., & Majumder, S. (2011, December). Shape, texture and local movement hand gesture features for indian sign language recognition. In *3rd International Conference on Trendz in Information Sciences & Computing (TISC2011)* (pp. 30-35). IEEE.

Rishabh, S., Raj, S., Nutan, V. B., & Prachi, R. R. (2016). Interactive projector screen with hand detection using gestures. In *International conference on automatic control and dynamic optimization techniques ICACDOT*. IEEE.

Samata, M., & Kinage, K.S. (2015). Study on hand gesture recognition. *International Journal of Computer Science and Mobile Computing, 4*, 51-57.

Schultz, M., Gill, J., Zubairi, S., Huber, R., & Gordin, F. (2003). Bacterial contamination of computer keyboards in a teaching hospital. *Infection Control and Hospital Epidemiology*, *4*(24), 302–303.

Siji Rani, S., Dhrisya, K. J., & Ahalyadas, M. (2017). International conference on advances in computing. In *Communications and informatics ICACCI*. IEEE.

Stefano, S., Paolo, M. R., David, A. F. G., Fabio, R., Rossana, T., Elisa, C., & Danilo, D. (2018). On-line event-driven hand gesture recognition based on surface electromyographic signals. In *International symposium on circuits and systems ISCAS*. IEEE.

Stergiopoulou, E., & Papamarkos, N. (2009). Hand gesture recognition using a neural network shape fitting technique. *Engineering Applications of Artificial Intelligence*, *22*(8), 1141–1158.

Sultana, A., & Rajapuspha, T. (2012). Vision based gesture recognition for alphabetical hand gestures using the svm classifier. *International Journal of Computer Science and Engineering Technology*, *3*(7), 218–223.

Tewari, A., Taetz, B., Grandidier, F., & Stricker, D. (2017, October). A probabilistic combination of CNN and RNN estimates for hand gesture based interaction in car. In *2017 IEEE International Symposium on Mixed and Augmented Reality (ISMAR-Adjunct)* (pp. 1-6). IEEE.

Vaibhavi, S.G., Akshay, A.K., Sanket, N.R., Vaishali, A.T., & Shabnam, S.S. (2014). A review of various gesture recognition techniques. *International Journal of Engineering and Computer Science, 3*, 8202-8206.

Vieira, A. W., Nascimento, E. R., Oliveira, G. L., Liu, Z., & Campos, M. F. (2012, September). Stop: Space-time occupancy patterns for 3d action recognition from depth map sequences. In *Iberoamerican congress on pattern recognition* (pp. 252–259). Springer.

Wang, J., Liu, Z., Chorowski, J., Chen, Z., & Wu, Y. (2012, October). Robust 3d action recognition with random occupancy patterns. In *European Conference on Computer Vision* (pp. 872-885). Springer.

Wu, Y., & Huang, T. S. (2001). Hand modeling, analysis and recognition. *IEEE Signal Processing Magazine, 18*(3), 51–60.

Xia, L., & Aggarwal, J. K. (2013). Spatio-temporal depth cuboid similarity feature for activity recognition using depth camera. In *Proceedings of the IEEE conference on computer vision and pattern recognition* (pp. 2834-2841). IEEE.

Yang, X., Zhang, C., & Tian, Y. (2012, October). Recognizing actions using depth motion maps-based histograms of oriented gradients. In *Proceedings of the 20th ACM international conference on Multimedia* (pp. 1057-1060). ACM.

Yang, Z., Li, Y., Chen, W., & Zheng, Y. (2012, July). Dynamic hand gesture recognition using hidden Markov models. In *2012 7th International Conference on Computer Science & Education (ICCSE)* (pp. 360-365). IEEE.

Ye, M., Zhang, Q., Wang, L., Zhu, J., Yang, R., & Gall, J. (2013). *A survey on human motion analysis from depth data*. Academic Press.

Yifan, Z., Congqi, C., Jian, C., & Hanqing, L. (2018). EgoGesture: A new dataset and benchmark for egocentric hand gesture recognition. *Transactions on Multimedia, 20*, 1038-1050. DOI . doi:10.1109/TMM.2018.2808769

Yoon, H. S., Soh, J., Bae, Y. J., & Yang, H. S. (2001). Hand gesture recognition using combined features of location, angle and velocity. *Pattern Recognition, 34*(7), 1491–1501.

Yuntao, M., Yuxuan, L., Ruiyang, J., Xingyang, Y., Raza, S., Samuel, W., & Ravi, V. (2017). *Hand gesture recognition with convolutional neural networks for the multimodal UAV control*. IEEE.

Zengshan, T., Jiacheng, W., Xiaolong, Y., & Mu, Z. (2018). WiCatch: A wi-fi based hand gesture recognition system access. *IEEE Access, 6*, 16911-16923.

Chapter 3
Hand Gesture Recognition:
The Road Ahead

Hitesh Kumar Sharma
University of Petroleum and Energy Studies, India

Tanupriya Choudhury
iD https://orcid.org/0000-0002-9826-2759
School of Computer Science, University of Petroleum and Energy Studies, India

ABSTRACT

Hand gesture recognition (HGR) is a natural kind of human-machine interaction that has been used in a variety of settings. In this chapter, the authors address work in the field of HGR applications in industrial robots, with an emphasis on processing stages and approaches in gesture-based human-robot interaction (HRI), which may be beneficial to other researchers. They examine several similar research in the field of HGR that use a variety of techniques, including sensor-based and vision-based approaches. They discovered that the 3D vision-based HGR technique is a hard but interesting study field after comparing the two methodologies. Hand gesture recognition has a major application for disabled peoples. The person who is not able to speak can use these kinds of systems to control the devices or various applications in his/her machine. This technology also has vast potential in the gaming industry. The gaming industry has already adapted this technique.

INTRODUCTION

Human Robert Interaction is such a field where hand gesture recognition can play a major role. Even though its implementation has already been introduced in Human

DOI: 10.4018/978-1-7998-9434-6.ch003

Robot Interaction but still some high advancements can be done. In other words, hand gesture recognition is a virtual form of sign language. Through hand gestures we can instruct the robot to perform various tasks (K. Banjarey et al. 2022).

For example: If a human wants a robot to make some coffee, there will be different hand gestures required for the same. The robot will scan the hand gesture, process it into its memory and will give the desired output. The desired output will be given after performing the following steps:

1. First hand gesture will instruct the robot to boil defined quantity of water.
2. Second hand gesture will instruct the robot to add fixed amount of coffee and sugar as prescribed by the human.
3. And with the final hand gesture the robot will serve the coffee.

Therefore, with the help of hand gesture recognition it is possible for the robot to perform every possible human day-to-day activities (K. Banjarey et al. (2021))(P. Pardhi et al. (2021)). In figure 1, we have shown the various applications of hand gesture recognition.

Figure 1. Applications of hand gesture recognition

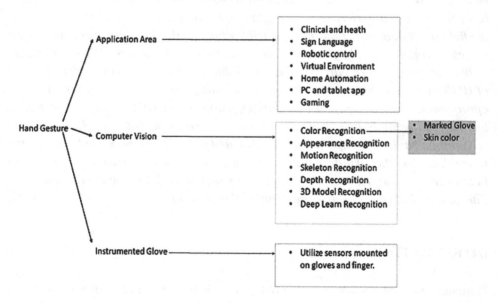

Today's youth is very fond of games. From children to adults everyone in today's time plays various types of games. Gaming along with its amazing graphics is a trend growing up so fast. We all have grown up playing board games like ludo king, snakes and ladders, monopoly, scrabble, axis and allies, chess and so on (Challa, S.K. et al. (2021)). In earlier time, these games were only present as board games but now these same games are present online and we can download them for free from Google Play Store. In 21st century, a game called as pubg is very famous. It is an online game and contains very amazing graphics and concepts of gaming (Dua, N. et al. (2021)). The only way to play pubg is to use your laptop/mobile screen along with your fingers giving instructions. What if we could use hand gesture recognition for the same? Hand gesture recognition would make the playing of games so simpler as by only moving your body parts in front of the screen would give computer instructions and the action will be performed inside the game. Hand gesture recognition will also benefit the human health as well. Right now, gamers are sitting on their chair and using only their fingers to play the game but with the help of hand gesture recognition gamer will be doing different actions which will allow the body to exercise. We can built a special software for the hand gesture recognition. For example: we can design an in-built feature inside the laptop/mobile cameras of gamers and insert the required gestures needed for the gaming and save it in the system memory. As peripheral devices, this system simply requires small, low-cost RGB (Red Green Blue) high speed cameras. At the same time, 3D modeling system based on hand gesture recognition also need to be introduced into the system (Dua, N. et al. (2021))(D. K. Dewangan et al. (2021)).

LITERATURE REVIEW

HGR is considered one of the most important feature of human body to identified his/her actual feelings and activities. An expert system used for automatic recognition of facial expression has lots of applications in this era. Animated Movies, Real-time advertisement, Healthcare, Gaming, Entertainment, sociable robotics and many others areas are the core potential sectors for these expert systems (Dua, N. et al. (2021)). Interaction between Human and Robot (HRI) is an emerging technology used for developing an interactional environments for a computational and expert system to understand emotional gestures and interact through some channels for communication between robot and human. It is used for responding to user based upon his/her emotional gesture (T. Mantecón et al. (2016)). Human creates many gesture on his/her face using eye, nose, mouth hands etc. (Kumar Sharma, H et al. (2018))(H.-J. Kim (2008)). There are many methods already discovered to identify these gestures and act accordingly. PCA (Principal Component Analysis)(Ali A et al. (2018)) and LDA (Linear Discriminant Analysis)(C. Szegedy et al. (2015))(HK

Sharma et al. 2013)(HK Sharma et al. (2015)) are some high accuracy methods have been used in many expert system for facial gesture recognition systems.

Hand gesture recognition plays an important role in medical sector. Based on the movements of hand, the wearable sensor device recognizes hand gestures. The concept is utilized by the Internet of Medical Things (IoMT). With its help it has been easier for the patients as well as the doctors to interact. IoMT has made things easier for the distanced and old patients who cannot visit the hospital on the regular basis. It work as a medium between the patients and the healthcare, by understanding the hand gesture the required service is provided to the distanced and old patients. Along with the medical assistance, IoMT also helps in maintaining the data of number of patients correctly and accurately. Data is stored digitally and this doesn't leads to paper wastage (D. K. Dewangan (2021)). For security purposes, fingerprints or pin locks can be used which also comes under hand gesture recognition. Doctors can detect the cause on their virtual screens. In coming future, if we gain more advancement on hand gesture recognition then it can be possible that surgeries will be performed through gestures. Hand gestures will be processed by the system and after processing the desired output/action will be performed, for example: picking up a tool to start a surgery. It will also enable distancing between the patient and the doctor, as during the pandemic it was risky for the doctors to go near the patient and do his/her check up because of the corona virus. With more advancement, assigning and billing of medicines to the patients can also be done with help of hand gesture recognition.

METHODS OF HAND GESTURE

Hand Gesture Recognition is one of the major application of Artificial Intelligence and Machine Learning. The Deep neural networks are used to train the system over a large numbers of hand images (Kumar Sharma, H et al. (2018)). There is a distinction made between tracking human movement and interpreting it as an important order. Gestures are often interpreted using one of two ways in HCI applications (H.-J. Kim et al. (2008)). Computer vision is used in the second technique, which does not require any sensors to be worn.

Glove-Based Hand Gestures

Glove-based sensors that capture hand motion and location can be used to aid in the development of new products. Hand gloves contains internal fitted sensors used to capture palm position, hand position, figure position and the orientation of the same gestures. This technique is used to generate the required dataset to train the model and for controlling the system after completing the training. In addition, these gadgets are expensive (Ali A et al. (2018))(C. Szegedy et al. (2015)). As a result of this new technology, contemporary gloves are regarded to be industrial-grade haptic technology. Use of a microfluidic glove to provide users with haptic feedback that lets them experience a virtual object's form, texture, movement, and weight.

Computer Vision-Based Hand Gestures

It is a widely utilized and flexible technology as a contactless communication channel. The employment of monocular, fisheye, TOF, or infrared cameras is conceivable. There are many drawbacks of this system. The background, the noise, the lighting effect, the orientation are some major challenges to capture the proper dataset. These reasons may cause for poor quality dataset.

Skin Color: Color-Based Recognition

A few of the numerous applications that use skin color detection for hand segmentation include object categorization and damaged picture recovery, human movement tracking and observation, HCI applications, facial recognition, and hand segmentation. With the help of two approaches, it has been possible to identify skin color. According to the first technique, each pixel in a picture is categorized as skin or not depending on its neighboring pixels. The second step is to process skin pixels spatially based on information such as intensity and texture. Mathematicians can make use of color space to express picture color information in their calculations. Multi-color space can be used for various application in TV transmission, Image processing and computer vision based systems. Variations in lighting, background noise, and other factors complicate the skin color technique. An HD camera was used to record 14 movements between 0.15 m and 0.20 m away as part of a study by primal and colleagues. Noise, light intensity, and hand size were utilized to determine the rate of recognition. While strong light during capture lowers accuracy, according to research by Suleyman et al., utilizing the Y–Cb–Cr color scheme is good for removing lighting effects. You can not use anything that matches the skin color as a backdrop! Choudhury et al. developed a novel hand segmentation approach that integrated frame differencing with skin color segmentation while dealing with circumstances that involve moving components in the backdrop, such as moving drapes and waving trees in the background (HK Sharma et al. (2013))(HK Sharma (2015)). Sotiropoulos et al. combined picture differencing and background removal with motion-based segmentation to achieve a robust system that overcomes illumination and complex backdrop problems.

USING APPEARANCE TO IDENTIFY A PERSON

A hand may be modeled using this approach by extracting picture characteristics and comparing those retrieved from the input frames with those extracted from the input images. Without a preliminary segmentation procedure, the pixel intensities are used to compute the characteristics of the image. 2D picture attributes can be obtained with ease, allowing for real-time processing. This method is much simpler than 3D modeling method since it can be implemented faster (Mais Yasen et al. (2019)). Using this method, it can also recognize various skin tones. Both the motion model and the static model can be used.

Using motion-based recognition, a series of image frames may be utilized to extract a specific item. The AdaBoost algorithm is required to recognize the gesture. Motion recognition suffers a setback if more than one gesture is used during the recognition process. Starting with the hand identified in each picture and its tension force, the tracking process begins. To improve classification of gestures, a second stage matching model is developed using the motion tracking. It was shown that could create a user-centric coordinate frame utilizing a conventional face identification method and optical flow computation to recognize gestures using motion characteristics and the multiclass boosting approach.

A segmentation technique was not required to classify the hand motions in the real dataset video. Aside from the TOF camera's depth range restriction, the system produces good results. Background and skin color were distinguished using the CAM Shift method in the In, YUV color space, and gesture recognition was assisted by the naive Bayes classifier. In addition to illumination variations, the skin segment results can be affected by changes in lighting conditions. Changes in the rotation have an impact on the output result. In figure 3, we have shown the techniques and challenges in hand gesture recognition.

THE RECOGNITION OF SKELETONS

It specifies model parameters that can help enhance the identification of complicated characteristics. Because it provides geometric properties and constraints as well as simple translations of features and correlations of data, it may be utilized for the classification of the hand model. This allows the user to focus on geometric and statistical information. Joint orientation, joint spacing, joint-position of skeletal, joints angle, and joint trajectory curvature are the mainly used features.

Kinect's depth sensor is used for hand segmentation, followed by 3D connections and Euclidean and geodesic distances over hand skeleton pixels for fingertip localization. Using parallel convolutional neural networks (CNNs) to analyze hand skeletal joint positions, it employs a deep learning model to analyze joint locations (Nico Zengeler et al. (2019))(Bhushan, A et al. (2017)).

SYSTEM HAND GESTURE RECOGNITION APPLICATION AREAS

As a result of hand gesture research, data gloves are less expensive to use, and they provide a more natural way to communicate. Gaming and industrial control consoles are examples of traditional interactive techniques. Some main features of hand gesture recognition is explained in following sections.

Several actions may be performed with these motions without needing a mouse, keyboard, or touch screen. Hand gestures can also be used to operate a wheelchair. It can also used in controlling a gaming character movements using hand orientation and its movement. Education sector is also having the major application of this approach. In figure 2, we have shown various systems controlled using HGR.

Figure 2. Hand gesture recognition based control systems

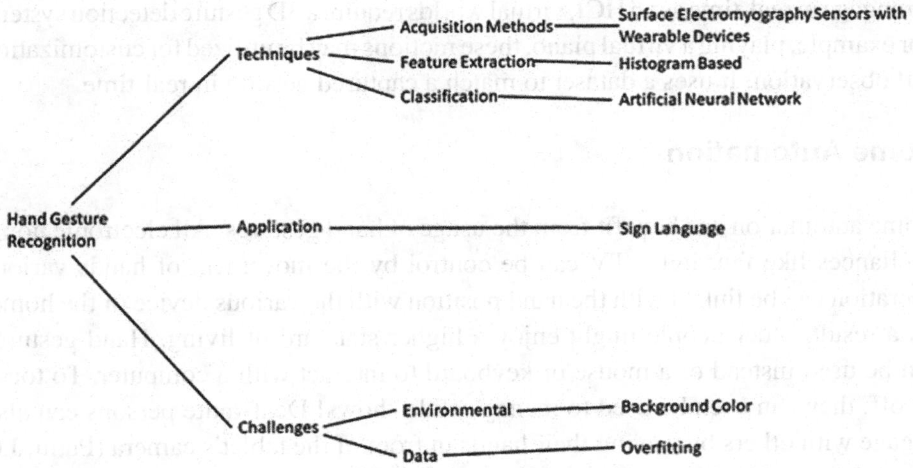

SYSTEM FOR RECOGNIZING SIGN LANGUAGE

People who are unable to communicate verbally utilize sign language as an alternate option. A series of hand movements are used to indicate a letter, number, or another statement. Many research articles have proposed that deaf-mute persons be able to recognize sign language by wearing a glove-attached sensor that responds to hand movements. We can also use computer vision technique to recognize the real time image. The real time captured images will be compared with the stored images and the computer vision techniques identify the posture and take necessary controlling action.

Automated Robotics

Many industries, including manufacturing, assistive services, retailing, sports, and entertainment, employ robot technology. Artificial intelligence, machine learning techniques, and complicated algorithms are used by robotic control systems to accomplish a given job. This allows the robotic system to interact organically with its surroundings and make an autonomous choice. Research suggests using computer vision technology and a robot in the development of geriatric assistance technologies (S Taneja et al. (2018)). While in another study, robots ask humans for directions inside certain buildings using computer vision.

Virtual Environments

To engage in real-time as an HCI, virtual worlds require a 3D gesture detection system. For example, playing a virtual piano, these motions may be utilized for customization and observation. It uses a dataset to match a captured gesture in real-time.

Home Automation

Home automation can benefit from the usage of hand gestures. All electronic home appliances like fan, light, TV can be control by the movement of hands various operation can be linked with the hand position with the various device of the home. As a result, older people might enjoy a higher standard of living. Hand gestures can be used instead of a mouse or keyboard to interact with a computer. To top it all off, they can even be used to manage slide shows! Deaf-mute persons can also engage with others by waving their hands in front of the tablet's camera (Patni, J.C et al. (2019)). You'll need to install an application on your computer to translate the signs into text, which is shown on your screen. As with the transfer of acquired speech to text, this is similar.

Gestures for Playing Video Games

Microsoft's Kinect Xbox is the best illustration of how gestures may be used to enhance gaming experiences in general. You can connect to the Xbox through a cable connector, and there's an above-the-screen camera. Users may interact with the game via hand and body movements thanks to Kinect's camera sensor (S Taneja et al. (2019)).

Metaverse

Metaverse too is a brand-new concept which means a virtual universe. One can create avatars and explore the metaverse using AR/VR and it can be the next big thing in social media. It can completely change the way we view the world and to transverse this metaverse HGR is an important piece of technology since we cannot use mouse/keyboards.

Holograms

Holograms are nothing but a 3D representation of images. But since they are a 3D representation, we need a way to manipulate them and that is where HGR comes in. Using various motions, we can change the size, orientation, colors etc. of the image.

Medical Simulators (Pre-Surgery Precision)

HGR is used in simulators to visualize precise components like heart valve and liver ducts before operations for surgeons to understand what they are dealing with. This is done by making 3D representation of the patient's scans.

HAND GESTURE RECOGNITION PROCESS

Hand gesture recognition (HGR) may be a natural means of Human Machine Interaction and has been applied on completely different areas aspects of life. during this paper, we'll discuss work exhausted the world of applications of HGR in industrial robots wherever focus is on the process steps and techniques in gesture-based Human golem Interaction (HRI), which may give helpful info for different researchers. we have a tendency to review many connected works within the space of HGR supported completely different approaches as well as sensing element based mostly approach and vision approach. once scrutiny the 2 approaches, we have a tendency to found that 3D vision-based HGR technique may be a difficult however promising researching space. Then, regarding works of implementation of HGR in industrial situation are mentioned well. Pattern recognition algorithms that effectively employed in HGR like k-means, DTW etc. are in short introduced still. In figure 4, we have shown various hand gesture images. (Figure 3).

Figure 3. Hand gesture images

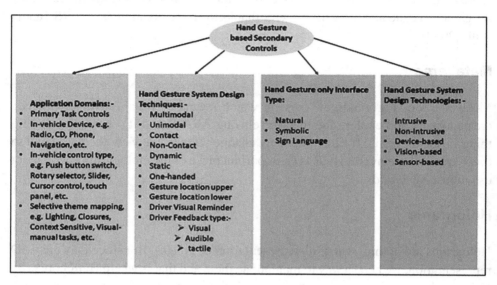

The computer or any electronic devise understands the given commands as hand actions through some of steps.

1. The user performs the hand gestures that represents the varied commands already grasp to the computer
2. The camera catches the image shaped by the user and analyses the amount of fingers shown and therefore the variety of type the hand has created to present a selected command
3. After understanding the hand gesture the pc starts functioning on the task commanded
4. Once all the inner messages ar send to the varied elements of the pc stating the tasks it's to perform, the machine provides the output and therefore the task commanded by the user is completed.

In figure 4, steps followed by HGR based system is shown.

Figure 4. The steps of hand gesture recognition by ADP system

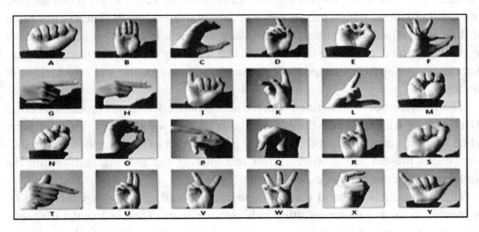

The hand signal is classified into many classes like static and dynamic or bodily property and gestures, etc. with every image having a unique which means there would be big selection of symbols that might simply be performed and therefore the human -computer interaction might be created a lot of less advanced and simple. In figure 5, we have shown the basic steps used by the HGR based system to control commands.

Figure 5. Hand gesture based command control system

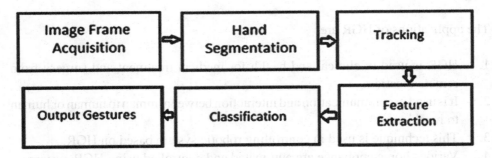

The first basic step is to create a list of commands that are to be defined. A neural engine is trained to recognize these commands. The image is to first be loaded and the resized. Following this, Hue Saturation Value is used to differentiate the hand from the background. The HSV value for the skin is (316, 93, 36). The margins of the hand are defined, and the fingers and palm are segregated, and thus the hand is extracted from the background. Stereo Cameras are also used to get more datasets and hand positions which can increase accuracy tremendously. The central point of the palm is used as a reference to measure distance. Each finger is individually recognized, and data sets are created. This is done using the depth sensing cameras. Accelerometer and Gyroscope track the motions, movements, and orientations of the hand.

Then a ML model is generated with the neural network that was processed above. After a lot of trial and error the ML model is trained for accuracy. The computer when adept at recognizing motions is the programmed to send various commands for various pre-defined motions. After recognizing the motion, and getting the command, the command is executed.

The recognition of hand can be done using wired gloves too which can be connected to the computer resulting in higher accuracy. They utilize the magnetic sensors, gyroscopes, force resistive sensors, inertial sensors, and electromyography.

HAND GESTURE RECOGNITION APPLICATION AREAS AND CHALLENGES

The current and future applications of Hand Gesture Recognition has shown in following figure (Figure 6). The major challenges faced by this approach is also shown in the same figure.

Applications

The applications of HGR are:

1. HGR is used in clinical and health for medical treatment and surgery from remote location.
2. It is used in communication and interaction between human to human or human to machine.
3. This technique is used in controlling robotic system based on HGR.
4. Various home appliance are automated and controlled using HGR system.
5. Gaming console, Desktop and laptops are also using HGR technique to access them smoothly.

Figure 6. Hand gesture recognition techniques and challenges

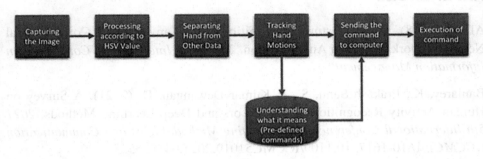

Challenges

The major challenges in using HGR are:

1. **Environment Compatibility:** The background color, texture and other intensity noise can affect the performance of hand recognized system.
2. **Training Dataset:** The training dataset can also be a major challenge in these systems. If the data is over fitted or under fitted, the performance may degrade of these system.

To overcome these challenges, these issues should be taken care in training these systems.

CONCLUSION

Recognizing hand gestures solves one of the major problems with interface systems. It is more natural, easier, more flexible, and less expensive to control things with your hands, and there is no need to address problems created by hardware devices, as none are necessary. To overcome common challenges and achieve a reliable result, it is important to create trustworthy and robust algorithms with the help of a camera sensor that has a certain feature, which needs substantial work. To be sure, each of the techniques listed above is not without its own set of pros and downsides.

REFERENCES

Alani. (2018). Hand Gesture Recognition Using an Adapted Convolutional Neural Network with Data Augmentation. *24th IEEE International Conference on Information Management.*

Banjarey, K., Prakash Sahu, S., & Kumar Dewangan, D. (2021). A Survey on Human Activity Recognition using Sensors and Deep Learning Methods. *2021 5th International Conference on Computing Methodologies and Communication (ICCMC),* 1610-1617. 10.1109/ICCMC51019.2021.9418255

Banjarey, K., Sahu, S. P., & Dewangan, D. K. (2022). Human Activity Recognition Using 1D Convolutional Neural Network. In S. Shakya, V. E. Balas, S. Kamolphiwong, & K. L. Du (Eds.), *Sentimental Analysis and Deep Learning. Advances in Intelligent Systems and Computing* (Vol. 1408). Springer. doi:10.1007/978-981-16-5157-1_54

Bhushan, A., Rastogi, P., Sharma, H. K., & Ahmed, M. E. (2017). I/O and memory management: Two keys for tuning RDBMS. *Proceedings on 2016 2nd International Conference on Next Generation Computing Technologies, NGCT 2016,* 208–214.

Challa, S. K., Kumar, A., & Semwal, V. B. (2021). A multibranch CNN-BiLSTM model for human activity recognition using wearable sensor data. *The Visual Computer.* Advance online publication. doi:10.100700371-021-02283-3

Dewangan & Sahu. (2021). Driving Behavior Analysis of Intelligent Vehicle System for Lane Detection Using Vision-Sensor. *IEEE Sensors Journal, 21*(5), 6367-6375. . doi:10.1109/JSEN.2020.3037340

Dua, N., Singh, S. N., & Semwal, V. B. (2021). Multi-input CNN-GRU based human activity recognition using wearable sensors. *Computing, 103*(7), 1461–1478. doi:10.100700607-021-00928-8

Kim, L., & Park. (2008). Dynamic hand gesture recognition using a CNN model with 3D receptive fields. *International Conference on Neural Networks and Signal Processing.*

Krizhevsky, Sutskever, & Hinton. (2012). *Imagenet classification with deep convolutional neural networks.* NIPS.

Kumar Sharma, H., & Kshitiz, K. (2018). *NLP and Machine Learning Techniques for Detecting Insulting Comments on Social Networking Platforms.* ICACCE.

Mantecón, T., del Blanco, C. R., Jaureguizar, F., & García, N. (2016). Hand Gesture Recognition using Infrared Imagery Provided by Leap Motion Controller. *Int. Conf. on Advanced Concepts for Intelligent Vision Systems, ACIVS 2016*, 47-57. 10.1007/978-3-319-48680-2_5

Pardhi, P., Yadav, K., Shrivastav, S., Sahu, S. P., & Kumar Dewangan, D. (2021). Vehicle Motion Prediction for Autonomous Navigation system Using 3 Dimensional Convolutional Neural Network. *2021 5th International Conference on Computing Methodologies and Communication (ICCMC)*, 1322-1329. 10.1109/ICCMC51019.2021.9418449

Patni, J. C., & Sharma, H. K. (2019). Air Quality Prediction using Artificial Neural Networks. *2019 International Conference on Automation, Computational and Technology Management, ICACTM 2019*, 568–572. 10.1109/ICACTM.2019.8776774

Sharma, Shastri, & Biswas. (2013). A Framework for Automated Database TuningUsing Dynamic SGA Parameters and Basic Operating System Utilities. *Database System Journal*.

Sharma, Shastri, & Biswas. (2015). Auto-selection and management of dynamic SGA parameters in RDBMS. *Database System Journal*.

Szegedy, L., Jia, S., Reed, A., & Erhan, V., & Rabinovich. (2015). Going deeper with convolutions. *2015 IEEE Conference on Computer Vision and Pattern Recognition (CVPR)*.

Taneja, S., Karthik, M., Shukla, M., & Sharma, H. K. (2018). Architecture of IOT based Real Time Tracking System. *International Journal of Innovations & Advancement in Computer Science*, 6(12).

Taneja, S., Karthik, M., Shukla, M., & Sharma, H. K. (2017). *AirBits: A Web Application Development using Microsoft Azure*. International Conference on Recent Developments in Science, Technology, Humanities and Management (ICRDSTHM-17), Kuala Lumpur, Malasyia.

Yasen, M., & Jusoh, S. (2019, September). A systematic review on hand gesture recognition techniques, challenges and applications. *PeerJ. Computer Science*, 5, e218. doi:10.7717/peerj-cs.218 PMID:33816871

Zengeler, N., Kopinski, T., & Handmann, U. (2019). Hand Gesture Recognition in Automotive Human–Machine Interaction Using Depth Cameras. *Sensors (Basel)*, 19(1), 59. doi:10.339019010059 PMID:30586882

Chapter 4
IoT–Based Wearable Sensors:
Hand Gesture Recognition

Bikram Pratim Bhuyan
School of Computer Science, University of Petroleum and Energy Studies, India

Shelly Garg
School of Computer Science, University of Petroleum and Energy Studies, India

ABSTRACT

Recognition of gestures has gained a lot of popularity in recent years and in the world-wide community as it has played an important role in many of the practical applications. A few of the popular practical applications are reading sign language, authentication of user, assistance provided in day-to-day life. Just to take an example, humans interact with small computers like mobile, wearable smart devices, popularly known as internet of things (IoT), which have a limitation to be embedded with in screens or keyboards. As per statistics, 26% of the population is under the age of 15 whereas some 9% of the population is above the age of 65 where they have problems performing a few activities on a daily basis, and hand gesture recognition can prove to be a good solution.

INTRODUCTION

Applications such as a natural human computer, omnipresent calculation, sign language, user identification and everyday help have substantially drawn the research community into attention. The research community has a wide array of practical applications. For example, interaction of wearable or Internet of Things (IoT) computers between human and small, mobile computers that cannot be integrated

DOI: 10.4018/978-1-7998-9434-6.ch004

with displays or keyboards. This leads to gestures by the user being one of the effective techniques for interacting human and intelligent gadgets. In addition, the interaction with human actions might enable the user to operate in a natural and unobtrusive manner and communicate with gadgets.

Typically, sign recognition using ambient sensors or wearable sensors can be approached. The first analyses a series of sensory data (i.e., pictures) taken from the sensors integrated into the environment, while the latter uses the sensor signals from devices such as intelligent clocks, intelligent gloves and mobile devices. Digital cameras (i.e. computer view), light and/or accelerometers and/or gyroscopes are the most frequent sensors used to recognise gestures and often are integrated into wearable and mobile devices. The advantages of the gesture recognition technique based on environmental sensors can be beneficial for home interaction and control. The ambient-sensor-based system nevertheless restricts the recognition of human motions in pre-definition contexts like a room and an apartment, whereas it can't be recognised anywhere at all time. Furthermore, environmental vision might increase users' privacy concerns. Wearable sensors, however, allow motions to be detected everywhere, while wearables may relieve the user's privacy problems.

IoT devices have recently been ever omnipresent in our lives, as they are cheaper, integrated with microprocessors, memory and cache and wi-fi, bluetooth and several capabilities such as microphone, accelerometer, gyroscope, magnetometer, light, theme, etc. This study examines the issue of continually detecting human motions using low-cost wearable IoT devices through the utilization of these technologies. The emergence of wireless sensing technologies has brought significant attention from researchers in HCI to gesture detection and is gradually replacing conventional input devices (e.g., keyboards and handles). The computers can comprehend and do certain activities (eg writing words and playing virtual reality games) inherent in human gesture commands.

This chapter starts with the various application of IoT in Gesture Recognition, it then ventures into the various datasets available for hand gesture recognition before dwelling inside the background study behind the same. Studies forwarded for the usage of machine learning and deep learning are also focused before the usage of ontology for knowledge representation and reasoning in hand gesture recognition.

IOT IN GESTURE RECOGNITION

Ambient sensors and wearable mobile sensors are two primary techniques for gesture recognition. Sensors like digital depth cameras or light sensors are preset in the environment in the ambient sensor method. Machine learning models are used to evaluate sensory input, for example image sequence or the user's shadow patterns.

The study (Doan et al., 2017) addresses the issue of the cyclical patterns of moving hand gestures using a space - time representation scheme, which takes into consideration the process of recognition both of the hand forms and their motions throughout a gesture. The method is verified by data sets of versatile, non-cyclical and cyclical hand gestures and has shown the 96 percent accuracy of cyclical patterns, which is therefore highly promising for interaction applications in biomedical application such as hand gestures for control of household appliance nonverbal cues.

The study (Venkatnarayan & Shahzad, 2018) used a range of light sensors in the house, named LiGest, an agnostic system for the user's illumination, location and orientation. LiGest uses the user's shadows in unique patterns and varied motions. The study also showed how the intensity, the size and the number of shadows of a user may be resolved since they depend on the position and orientation, intensity and position as well as the number of light sources.

Substantial progress has been achieved in recognition of gestures. However, some limits on the usage of ambient sensors cannot be used everywhere to recognise motions in a room or apartment. And environmental sensors frequently create the user's privacy problems as they are typically put in the private area.

During his studio, Kim (Kim et al., 2012) created a wrist-wearing instrument which recovers the user's entire 3D position and allows for a range of freelance interactions in motion. Applications such as 3D spatial engagement with mobile devices and eye-free interaction on the move and gaming can be demonstrated in the study.

The study (Pham, 2015) has leveraged the sensors inserted within the mobile device and offers a set of characteristics invariant of the device rotation and position change, which are derived from the constant sensing data stream, to recognise human motions and actions in diverse settings. Due to the rapid growth of mobile technology, wearable and mobile sensors become inexpensive and strong. In this study, we utilise a low-cost and low-battery IoT device capable of capturing different patterns of the user's movements constantly.

Management recognition automates the detection job of human actions in a sensing environment based on devices or without devices. The task of recognition uses advances in wireless technology for the detection and recognition, depending on the range of wireless communication protocols, of human goals in an indoor or indoor environment. State-of-the-art equipment-free sensing employs electromagnetic goods or commercial off- The shelf (COTS) of electromagnetic spectrum products. Device-free RF Sensing techniques utilize several IoT connection protocols and overcome the aforesaid constraints by setting a contactless recognition paradigm. Radio frequency sensors that are used for COTS devices' wireless signals conduct a non-intrusive and non-obtrusive activity recognition that operates within a variety of frequencies and which allows the recognition task to take place according to its coverage and spectral efficiency.

The sensor type, signal processing methods and classification algorithms utilised for the amount of identified motions are accurately stated. The sensor or video application may be found to pre-process the obtained signal or image data and provide it as an input to the study methods. Although these devices identify well, the use of such devices poses a comfort problem and threat to the individual's privacy. The recognition performance diminishes owing to signal interferences with more than one participant on the sensing region. The radar signals contain background noise and are thus subject to techniques of pre-processing or signal modification. The choice of the implemented learning algorithm depends on the data gathering and signal processing methods for the classification problem. But in reality it is difficult and not adapted for all interior settings to deploy such specialized hardware.

HAND GESTURE RECOGNITION USING VISION

Tom Cruise switches on a digital screen on a wall, merely by lifting his hands wrapped in black wireless gloves on the future of Steven Spielberg's Minority Report. Like the director of an orchestra, he acts empty, attempting to play, magnifying and pulling video with his hand movement and his wrist twists. In the year 2054, the Minority Report will take place as shown. It illustrates the touchless technology that may be produced many decades before the attention Vision-based interfaces have garnered in recent years is visible.

Manipulation is a strong tool for human communication. Actually, gestures are so firmly ingrained in human communication that when individuals talk on the phone, they often keep on gesturing. The mode of speaking for communicating thoughts is supplemented by hand gestures. Hand gestures in a discussion relate to graduation, the organisation of the speech, the structure of the space and time. Thus, it may be accomplished via manual gesture for communication between people and computing systems.

The main question in the interplay of gestures is how computers understand the motions of hands. The existing techniques may be split mostly into data-guide approaches and vision approaches. Data glove-based approaches employ sensor instruments for digitalisation into many parametric data of hand and finger movements. The additional sensors make setup and movement easier to gather. The gadgets are nonetheless extremely costly and provide consumers with a very lengthy experience (Schiphorst et al., 2005). In contrast, only a camera is needed for Vision-based approaches (Quek, 1994), which allows for a natural connection between people and computers without additional equipment. These systems are designed to supplement biological view by defining artificial vision systems in software and/ or hardware. This is a challenge as these systems have to be unchanged, insensitive to illumination, person and camera to achieve real time performance. Furthermore, systems such as accuracy and resilience need to be improved to match the demands.

The hand movement is captured by the video camera in visual hand gesture recognition system (s). This movie is broken down into a range of features that take into consideration individual frames. Some filtering can also be carried out on the frames to remove superfluous data and to emphasize the components needed. The hands are, for example, separated from other portions of the body and from other things. For different positions the isolated hands are identified. Gestures may be learned against a hypothetical grammar, as they are nothing than a succession of hand position linked by continuous movements.

Appearance-based methods employ picture characteristics to model the hand's visual appearance and compare them to the image characteristics retrieved from the video entry. In general, appearance-based methods offer the benefits of real-time performance because of the simpler 2-D picture characteristics. The concept underlying the invariant characteristics is that an item may be represented as an assembly of these areas, i.e. instead of representing the thing in its whole, as a collection of typical components if it is feasible to identify distinctive spots or regions on objects. This has the benefit that partial occlusions and significant deformations of the point of view of an object may be readily managed. The item can still be discovered, provided that sufficient number of typical areas can be recognized. Therefore, for real time hand detection, these techniques look rather promising.

Recognition of hand gestures by vision remains a key research field since the algorithms available in comparison of mammalian vision are very rudimentary. A major difficulty which hampers most methods is that they have numerous underlying assumptions that are not universal to arbitrary conditions but may be appropriate in the controlled laboratory environment. Various typical assumptions include: stationary settings and ambient illumination conditions presuming strong contrast. The recognition findings published in the literature are also reliant on the gathering of data by each author, making it hard to compare techniques and also increasing the suspicion of universal applicability. In addition, most techniques have a feature set that is limited. The current trend for identification of hand gestures is the use of AI in training classifiers, however the training process generally involves a lot of data and it is a time consuming operation to choose the characteristics characterizing the item being recognized.

Another difficulty remaining is the acknowledgement of the point of time of beginning and finish of significant hand gestures (s). Sometimes this issue is known as "gesture spotting" or the segmentation of temporal gestures. We have great but fascinating research challenges to achieve strategies to minimize the duration of training and build a cost-effective real time signing system, which is resilient to environmental and light conditions and requires no additional hardware.

Static recognition always employs linear and non-linear classifiers. Time delay neural network, dynamic time warping, Finite State Machine, Hidden Markov (HMM), and common approaches Convolutionary neural network are the techniques for dynamic recognition of dynamic gestures (CNN). It is observed that HMM, SVM, CNN, etc. are the most often utilized techniques of recognition (Trigueiros et al., 2014). The hand recognition should satisfy three requirements, including strength, accuracy and real-time. Previous static gestures are comparable to placing motions on a plane, which may communicate extremely little meaning such as "OK" and "numbers."

Deep gesture information may be collected with the advancement of camera and sensor technologies. Researchers focus more on dynamic recognition of gestures. In a period of time, a sequence of activities concentrate on dynamic gestures to enhance temporal and dynamic information. The combination of these basic and static gestures is the dynamic recognition of gestures to communicate its combination. The 3D recognition technology is more precise and deeper than two-dimensional with the development of equipment.

A feed-back neural network called the Convolutional Neural Network is now the most popular approach used to recognise gestures. CNN's artificial neurons may react to part of the peripheral units and are effective in the processing of big images. However, CNN needs many photos for training or reusing portion of the large data-trained neural network. However, CapsNet utilises far fewer generalising training data. CNN is not able to handle very effectively with fluidity.

A network of capsules is composed of a small group of neurons. In a specified portion of the picture a capsules can learn to look at a certain item (Sabour et al., 2017). In the CapNet data are recorded in the network without initial loss and recovery on attitudes such precise position of the item, rotation, spacing, inclination, size, etc. Slight change in the input will bring little change to the output and data is preserved, allowing CapsNet to handle many visual tasks using a simple and consistent design.

Visually based are the majority of mature applications for human-machine interaction. Wearable interactions are still on the screen. Because wearable technology is not sufficiently developed, only a few companies are producing wearable gesture recognition devices. In future, the material will be adequate to allow people to wear these gadgets without a sensation that they do not have any impact on creating a gesture if major advancements are made in chip, material and battery technology. Therefore, contact-based identification of human hand gestures will have greater study support. Regarding the current scenario, the recognition of vision-based hand signs gives greater value to research, as well as additional goods and applications for research assistance. Now we guide our article to some machine learning techniques.

Machine Learning in Hand Gesture Recognition

Without being explicitly designed, machine learning is an application of artificial intelligence (AI) that enables systems to automatically learn and improve based on their experience. With machine learning, computer algorithms can access data and learn on their own. It is a field of artificial intelligence and computer science that leverages data and algorithms to mimic the way people learn, progressively increasing its precision. It is widely used for problems like as classification, identification, detection, and prediction. Automating data-driven procedures is also highly efficient. An output-generating model can be created using data. A fresh input may yield a correct answer, or the output may be able to make predictions about the data that is known. As a machine learning approach, deep learning belongs to a larger family. A mathematical model is created using layers that process the incoming data, extracting features.

The amount of contact between humans and computers has risen dramatically, and the field is always evolving, with new methods and strategies being developed. As a result of computer vision and artificial intelligence, hand gesture recognition has become one of the most sophisticated areas in which communication with deaf people may be improved as well as gesture-based intercellular communication. This real-time hand gesture recognition tool has as its primary goal the classification and recognition of gestures. To comprehend the movement of a hand, we employ diverse methods and concepts from a variety of approaches, such as image processing and artificial neural networks. Hand gesture recognition has a wide variety of applications. For example, we may converse with deaf persons who are unable to hear using their familiar sign language. There are a multitude of computer vision algorithms that could be used to identify and determine what type of gesture is really being sought by each of the methods. Hand gesture recognition is a growing field, and there are a number of implementations that use both machine learning and deep learning approaches to detect a gesture that is intoned by a human hand. To further comprehend the mechanics of the hand gesture recognition approach, a few publications are examined. Various researches have been performed out on the interpretation of gestures as the topic expands and there are different systems which include machines and techniques of profound learning to identify a human hand-intonated motion. Some articles are also investigated to explore the rationale for comprehension of hand gestures.

The article (Chen et al., 2020) revealed that one of the most used and famous architectures, CNN, achieved greater rates of component perception efficiently at minimal computer costs. The strategy only focused on the instances of gestures that existed without a hand place in a static plot and the manual impediment was followed with 24 gestures. The backpropagation algorithm and segments algorithm were used in (Colli-Alfaro et al., 2019) to prepare the multi-layer propagation and to determine the effect of backpropagation. It used a deep convolution topology, and filtering included a variety of geographical segmentation and identification.

There is a prominent approach used in several additional detecting applications among these techniques and methodologies, Hidden Markov Models (HMM) (Elmezain et al., 2009). There are several detection variations commonly employed by the programmer that is found, and the application referred to in this article refers to inspections and works with them all by learning and looking at other documents. The three media: images, video and webcam are addressed in this application. The detection process is via video and employs the HMM, using a general paper produced by Francois which refers to an application which identifies posture. This application is linked to the method undertaken by Francois and other researchers, which is firstly referred to by extracting information from either image, video or webcams in real time. The three approaches described above identify these characteristics, but the primary issue of all methods, whether connected to CNN, RNN or utilizing any other technically, is the use of all the ways of fitting and this refers to the bounding box presented in this work. The box reflects the detected information from which they acquire trust and the greatest value is the output of the image shown. In addition, all additional information differs depending on the approach used by others; nevertheless, certain other devices and techniques linked to separation, general location and even fusion of other partials aid the detection and recognition tasks attained.

Recognition of social touch by CNN proposed in (Nyirarugira et al., 2013). He used different computations, such as random forests (RFs), algorithms boosting and the Decision Tree algorithm, to detect movements, to use 600 palmar picture arrangements with 30 instances using the convolutionary neural organization, and to discover the best approach in a Network 8/8 framework. The length of the case is nonetheless flexible such that the result determines the optimal length of the case. It is done with a late assessment utilizing a data set with diverse people who execute social changes. A system that collects touch gestures with a deep neural organisation in an almost continuous way is good at present. The results demonstrated that their method worked better when differentiated, while the prior work was based on cross-subscription of the CoST dataset for the one-subject. The method is distinguished between two places of interest, which have 66.2 per cent accuracy in the perception of social acts.

In the research of multiscale CNNs for hand recognition in the area of vision, various approaches for hand differentiating evidence were presented in the last decade, motivated by the progress of article recognition (Ju et al., 2019). The worst approach is the identification of the skin tone, which works on the hands, faces, and arms, but also has problems for spectacular alterations because of their impact. The components of this multidimensional nature combine a considerable impediment, poor objectivity, varying lighting conditions and different hand motions and an unbelievable coordinated effort between hands and dissents or different hands. They also introduced a multi-scale R-CNN method to properly identify the hands of humans in unconstrained photos. The CNN model is able to get preferential outcomes over conventional VGG16 by its merger of staggered overturning features, achieving virtually 85% of 5,500 test pictures and 5,500 for setup.

(Do et al., 2020) developed the Conv1D multi-level LSTM, Conv2D and LSTM block feature. Proposed Skeletal point-cloud functionality from skeletal data as well as profound characteristics from the hand part segmentation model were used by the suggested technique. The technique reached 96.07 and 94.40 percent accuracy in the 14- and 28-class Dynamic Hand Management Recognition (DHG) dataset. In the study, dynamic hand motions from 14 depths and 28 skeletal data were recovered from the LSTM model with two convolutionary pyramid blocks. The precision was 94.40 percent in 18 classifications.

Inanition of high-level gesture representations by utilising Convolutionary Residual Networks (ResNets) to learn the temporal characteristics of colour pictures and Convolutionary Short-Term Memory Networks (ConvLSTM) was developed in (Elboushaki et al., 2020). A 2D-ResNet-based two-stream architecture was subsequently developed to take deep elements from gestures.

Combination of a ConvLSTM network feature fusion network for the extraction of local, global and deep spatiotemporal feature information was studied in (Peng et al., 2020). Local information has been obtained from 3D residential films, while a dynamic gesture has been learnt by the ConvLSTM network from the global spacetime information. 95.59% on the Jester datasets and 99.65% on the SKIG (Sheffield Kinect Gesture) data package were achieved using the suggested method.

The article (Rahim et al., 2019) looked at the translation of a sign word motion into a text. The authors of this article conducted the segmentation skinMask to extract functionality along the CNN. The vector support (SVM) was used to accurately categorise the gestures of the signals with a dataset of 11 gestures from a single hand and 9 from double hands.

The research (Mambou et al., 2019) examined hand movements from interior and outdoor settings in night linked with sexual assault. The system of gesture recognition was introduced using the Yolo CNN architecture, which extracted motions from hands and a categorization of bounding box pictures, which ultimately created an attack alert. The network model in general was not lightweight and less accurate.

In short, a powerful hand gestural recognition frame that overcomes the most frequent issues with fewer constraints and produces precise and dependable results is the main difficulty facing the scientists. Effective hand motions processing also has some limits, such as lighting variations, backdrop issues, distance and difficulty with many movements. There are ways to hand gesture identification using non-machine learning algorithms, however there is an issue since the precision fluctuates and it overlaps a gesture with another in different contexts, making it less flexible and unable to adjust separately. Thus machine learning systems are the need of the hour.

ROLE OF DEEP NEURAL NETWORKS

Advances in human visual processing have demonstrated a hierarchical way of perception. The biological drive for deep learning is based on studies of the visual cortex, which shows that abstract features from the visual input are integrated in the second layer into main features. Those features are then combined in the next layer into better defined characteristics (Najafabadi et al., 2015).

Modeling difficulties with deep networks were handled using the traditional backpropagation learning technique to train these networks; numerous researchers shown that performance degrades as we stack more cached layers (Erhan et al., 2010). While overfitting is due to the increasing number of neurons layered by additional hidden layers. However, the relatively recent training of deep networks in a greedy way, with deep weights one level at a time in a phase known as "pre-training," has shown significant effectiveness (generative model). The whole network may be improved using the backpropagation learning method for a classification job (discriminative model). Convolutionary neural networks often have convolution layers, sub-samples and a classifier layer (e.g., SVM or multilayer network). The matching sub-sample layer follows every layer of the convolution. The co-sampling layer and the sub-sampling layer contain maps arranged in 2D topology; in these networks many layers can be layered to build a deep network.

Auto-encoders are synthetic networks to learn how to recreate input data from the output layer (pre-training). These networks employ an unattended approach since class labels are not used to train. In general, an auto-encoder may be seen as having a hidden layer, which learns how the input characteristics can be compressed and expanded to the original input data in the output layer. The encoding step is the learning of hidden neuron activations, whereas the decoding stage is the learning of neuron activations from the hidden neuronal activation of the already experienced layer. Input data is given in the input layer, the same input information is delivered to the output layer as the desired output; the network encoder weights and decoder weights are then obtained through cost optimization. The sum of the cross-entropies from Bernoulli may be utilized as the cost function for binary valued inputs.

The autoencoder can be finished using the backpropagation learning method with the class labels of the training data for classification jobs. In addition, more dispersed and hierarchical feature learning has been demonstrated by stacking hidden layers (Vincent et al., 2010).

Many studies are based on substantially or comparatively differentiable movements to recognize statically hand gestures, as it is evident that certain hand gestures in 2D appear fairly identical. Therefore, sub-sets of signals in the database for recognition are extracted. However, in all of the movements that people can notice, minor distinctive characteristics are present. It is of great importance that human-computer system may increase its vocabulary by recognizing more gestures, since this permits an increase in the level.

The time dimension in sequences is generally a challenge in recognising actions/gestures, both in terms of data quantity and in the context of model complexity - particularly important features to the training of large parametrical networks of deep learning. In this regard, the writers (Waibel et al., 1989) offered a number of methods, such frame sub-sampling, aggressively mid-level video representation of local frame-level characteristics or temporal sequence modelling, to mention just several. In the past, academics have been trying to harness recurrent neural networks (RNNs).

These models, however, generally confronted certain important mathematical problems (Bengio et al., 1994). Long short term memory (LSTM) was finally incorporated in (Hochreiter & Schmidhuber, 1997) for the RNNs. While the use of deep learning to recognize actions and gestures is relatively new, the quantity of research published on these subjects in recent years is amazing. In spite of this, no prior survey of all existing work on deep learning to recognize action and gestures is available to the best of our knowledge. This study seeks to take a picture of current trends, including a thorough review of several deep models with an emphasis on how they interpret the data's time dimension.

How to address a temporal component is the most important problem in profound human motion and gesture detection. We classify methods into three broad groups based on this (Baccouche et al., 2011). In a convolutionary layer, the first group employs 3D filters. The 3D convolution and 3D pooling in CNN layers can capture both spatial and temporal discriminatory characteristics while preserving a particular temporal structure. In the second category, movement characteristics such 2D optical flow maps and input to the networks are pre-combined. Harvested gesture capabilities can be transmitted to the network as extra channels for the appearance or as a secondary network entrance (subsequently in conjunction with the preceding).

Recurrent Network Recurrent Neural (RNN) has become one of the structures most frequently employed to accommodate time data in hidden layers utilising recurrent connections. The short-term memory of this network is insufficiently important for action in the actual world. Long Short Term Memory (LSTM) has been proposed to overcome this problem, and is generally employed as a hidden RNN layer. Some additional effective extensions of RNN in detecting human activity include bidirectional RNN (BRNN), Hierary RNN (H-RNN) and differential RNN (D-RNN). There are many other techniques for temporary modelling, such as HMM.

In general, while comparing the approaches, there are two key issues: how does a method deal with temporal information? And how can tiny datasets train such a huge network? As mentioned, 3D filters in their 3D convolution and pooling layers allow techniques to learn motion characteristics. 3D networks have been demonstrated to learn complicated temporal patterns over a lengthy span of time. The weight initialization problem was explored due to the needed amount of data. The conversion of 2D convolutionary weights to 3D models gives more precision than retraining slate. The use of training networks on preset motion features has also been demonstrated to be an effective means of preventing them from implicitly learning motion characteristics. In addition, it has shown useful to finalize motion-based spatial data networks (ImageNet). Allow networks that are fine-tuned to achieve high performance on packed optical flow frames, albeit the training data are restricted.

Nevertheless, only limited time information (local) may be used by both parties. The most important benefit of third group methods (i.e., temporal models such as RNN, LSTM) is that they have long term time relations.

The main purposes of these models are the skeleton information. When working with skeletal data, such models are recommended. Because of the low dimensional skeleton characteristics, these networks are less weighted and can thus be trained using less data. The efficiency depends on the amount of data, regardless of the model. The collectivity is currently trying to develop bigger data sets that can copy large parametric deep models and the challenge organisation, which may advance the state of the art and facilitate comparing deeper learning architectures, by creating new data sets and clearly defined evaluation procedures (Abu-El-Haija et al., 2016).

Data enhancement and pre-training strategies are prevalent. Training techniques have also been developed for avoiding overfitting (i.e. dropping off) and controlling learning rate (e.g. SGD extensions and Nesterov impetus). Taking the whole time scale into account, it results in a large number of learning weights. An excellent approach is to reduce spatial resolution while increasing the time length in order to tackle this challenge and minimize the weight.

Data fusion is another method to improve the outcome of deep models. Different porciones of input data, main characteristics, information points etc. may be taught by individual networks, and then merged. Together learning has demonstrated that learning algorithm has reduced the bias and variance mistakes. In that sense we uncover new approaches that combine profound models for recognition of actions and gestures. It is popular recently that in action/gesture recognition contests, such techniques may be used to get the most efficient performance by slightly improving the model. Code distribution amongst peers is felt as a major advantage of deep learning over the other methodologies. In upcoming applications/areas such as social signal processing, emotional computing and personality analyses, deep learning will dominate.

ONTOLOGICAL ANALYSIS OF IOT BASED HAND GESTURE RECOGNITION

Many recent research on the problem of the identification and inclusion of hand gestures into the design and development of gestural interfaces can be found (Challa et al., 2021; Dewangan & Sahu, 2021; Dua et al., 2021; Pandey et al., 2018; Sahu & Dewangan, 2021). In majority of these situations, the meaning and behaviors of gestures are predetermined. However, the research appear to examine the capacity to recognize the link beyond a preset gesture mapping. Therefore we find relatively few research trying to identify and formalize the link between each gesture. One method taken by researchers is to identify taxonomies that let designers and producers, when developing gestural vocabulary, to employ standard meanings.

After this a study, (Scoditti et al., 2011) suggested a taxonomy for gestural interactions, to lead designers and researchers, which "require a systematic general framework which helps them reason, compare, generate (and build) the right approaches to deal with the challenge." They intend to create uniform languages across the system with a special focus on gestures. The authors do not connect current gesture vocabulary with semantic relationships, however. This study, however, is confined to the six television (43 gestures) orders and blind(s) used in the trial. Finding the capability and applicability of the proposed taxonomy and notation approach necessitates more testing with a growing number of orders. This notation also employs a numerical language that, unless designers consult a reference book, is difficult to understand. Furthermore, they claim that the size or speed of the hands were not considered in their methodology.

Beyond taxonomies, there is already research utilizing ontologies. The researchers in (Ousmer et al., 2019) created a gesture ontology based on a Microsoft Kinect-based skeleton that seeks to characterise human body mid-air motions. Their ontology is primarily concerned with portraying the holistic posture of the human body, and so overlooks features such as finger stance or motions, as well as a comprehensive depiction of the hand. Furthermore, the ontology is not freely disclosed, which limits its usage and extension. Furthermore, their primary contribution is to create a "sensor-independent ontology of body-based contextual gestures, with intrinsic and extrinsic properties," where distinct gestures are mapped with their meaning. The paper performed a similar investigation with the goal of solving the problem of increasing the knowledge level of computational systems in order to identify gestural information with relation to arm motions.

In their research, they aimed to characterise knowledge of arm movements and to recognise them with greater precision. This is an intriguing research in which the authors employed Qualisys motion capture (MOCAP) to record the mobility of the participant's upper forearm while delivering an arm gesture. Their attention, however, was mostly on identifying geometrical movements, and the gesture set was limited to 5 geometrical forms. Again, their ontological framework does not take into account the mapping of other gestures with comparable referents.

The Human Device Gesture Interaction (HDGI) ontology is presented in this work (Perera et al., 2020), where a framework of human device gesture interactions that specifies gestures linked to human device interactions and translates them to relevant discourses. This is the first step in creating a comprehensive human device gesture interaction knowledge library, with the ultimate aim of improving user experience. The HDGI ontology can help motion detection systems, designers, manufacturers, and even developers formally articulate gestures and carry out automated reasoning tasks based on gesture and device affordance connections.

Large groups of Kinect-based motions, such as editing, grouping, distance-computer, aggregating, composition, and organizing, searching methods are managed in an integration-based fashion in order to recognize and develop automatically but not for managing them for human reasons. This implies the necessity to arrange such acts according to a consistent representation, which permits this reasoning with intrinsic (i.e. gesture-related) and extrinsic (i.e. user-related) characteristics.

Ontology Web Language has been encoded with the Ontology Web Language an ontology for organising body-based gestures based on user, body and body components, gestures and the surroundings in (Ousmer et al., 2019). A gesture elicitation research was carried out with 24 people who requested 456 body-based gestures for 19 referents linked with frequent IoT tasks, to demonstrate this ontology and to foster this ontology. The gestures were first requested and categorised according to criteria to identify 53 separate gestures in 23 categories, each of which might possibly have sub-categories.

CONCLUSION

This article comprehensively addresses the recognition of gesture without device that uses model-based and pedagogical methodologies. Appropriate means of acquisition and signalling or pre-conditioning as it contributes to performance must be robustly recognised. Due to its privacy preservation and its non-intrusive qualities, Wi-Fi CSI sensing was regarded more convenient than conventional techniques. Less sample size overfitting of learning-based systems demands an adequate signal prerequisite for enhanced recognition. The classification work of learning algorithms also delivers inferior performance with untrained or unseen material. Extensive study is focused on gesture recognition owing to the influence on the accuracy of signals, function extractions and selection approaches. The state-of-the-art processing of signals and extraction methods could only be observed with statistical moments of first and second order.

Choosing between a standard learning approach and a deep learning method is also influenced by the quantity of data gathered during the data collection stage. Deep learning relies on a large dataset that simultaneously extracts and classifies characteristics for excellent performance. As of yet, the connection between the occurrences and measurements of the discoveries reported is still being investigated. Learning-based approaches may be seen in a similar trend with large data trade volumes. Rather of relying on a large number of hand-made qualities, machine learning techniques consistently perform well. Even if there is a lot of literature on model-based and learning-based solutions, academics may still explore a scenario of multi-user engagement. We could employ hybrid approaches and signal information from numerous sensors to better carry out the recognition job.

A hand gesture recognition ontology survey is finally conducted.

REFERENCES

Abu-El-Haija, S., Kothari, N., Lee, J., Natsev, P., Toderici, G., Varadarajan, B., & Vijayanarasimhan, S. (2016). *Youtube-8m: A large-scale video classification benchmark.* arXiv preprint arXiv:1609.08675.

Baccouche, M., Mamalet, F., Wolf, C., Garcia, C., & Baskurt, A. (2011, November). Sequential deep learning for human action recognition. In *International workshop on human behavior understanding* (pp. 29-39). Springer. 10.1007/978-3-642-25446-8_4

Bengio, Y., Simard, P., & Frasconi, P. (1994). Learning long-term dependencies with gradient descent is difficult. *IEEE Transactions on Neural Networks*, 5(2), 157–166. doi:10.1109/72.279181 PMID:18267787

Challa, S. K., Kumar, A., & Semwal, V. B. (2021). A multibranch CNN-BiLSTM model for human activity recognition using wearable sensor data. *The Visual Computer*, 1–15. doi:10.100700371-021-02283-3

Chen, L., Fu, J., Wu, Y., Li, H., & Zheng, B. (2020). Hand gesture recognition using compact CNN via surface electromyography signals. *Sensors (Basel)*, 20(3), 672. doi:10.339020030672 PMID:31991849

Colli-Alfaro, J. G., Ibrahim, A., & Trejos, A. L. (2019, June). Design of user-independent hand gesture recognition using multilayer perceptron networks and sensor fusion techniques. In *2019 IEEE 16th International Conference on Rehabilitation Robotics (ICORR)* (pp. 1103-1108). IEEE. 10.1109/ICORR.2019.8779533

Dewangan, D. K., & Sahu, S. P. (2021). RCNet: Road classification convolutional neural networks for intelligent vehicle system. *Intelligent Service Robotics*, *14*(2), 199–214. doi:10.100711370-020-00343-6

Do, N. T., Kim, S. H., Yang, H. J., & Lee, G. S. (2020). Robust hand shape features for dynamic hand gesture recognition using multi-level feature LSTM. *Applied Sciences (Basel, Switzerland)*, *10*(18), 6293. doi:10.3390/app10186293

Doan, H. G., Vu, H., & Tran, T. H. (2017, May). Dynamic hand gesture recognition from cyclical hand pattern. In *2017 Fifteenth IAPR International Conference on Machine Vision Applications (MVA)* (pp. 97-100). IEEE. 10.23919/MVA.2017.7986799

Dua, N., Singh, S. N., & Semwal, V. B. (2021). Multi-input CNN-GRU based human activity recognition using wearable sensors. *Computing*, *103*(7), 1–18. doi:10.100700607-021-00928-8

Elboushaki, A., Hannane, R., Afdel, K., & Koutti, L. (2020). MultiD-CNN: A multi-dimensional feature learning approach based on deep convolutional networks for gesture recognition in RGB-D image sequences. *Expert Systems with Applications*, *139*, 112829. doi:10.1016/j.eswa.2019.112829

Elmezain, M., Al-Hamadi, A., Appenrodt, J., & Michaelis, B. (2009). A hidden markov model-based isolated and meaningful hand gesture recognition. *International Journal of Electrical Computing Systems in Engineering*, *3*(3), 156–163.

Erhan, D., Courville, A., Bengio, Y., & Vincent, P. (2010, March). Why does unsupervised pre-training help deep learning? In *Proceedings of the thirteenth international conference on artificial intelligence and statistics* (pp. 201-208). JMLR Workshop and Conference Proceedings.

Hochreiter, S., & Schmidhuber, J. (1997). Long short-term memory. *Neural Computation*, *9*(8), 1735–1780. doi:10.1162/neco.1997.9.8.1735 PMID:9377276

Ju, M., Luo, H., Wang, Z., Hui, B., & Chang, Z. (2019). The application of improved YOLO V3 in multi-scale target detection. *Applied Sciences (Basel, Switzerland)*, *9*(18), 3775. doi:10.3390/app9183775

Kim, D., Hilliges, O., Izadi, S., Butler, A. D., Chen, J., Oikonomidis, I., & Olivier, P. (2012, October). Digits: freehand 3D interactions anywhere using a wrist-worn gloveless sensor. In *Proceedings of the 25th annual ACM symposium on User interface software and technology* (pp. 167-176). 10.1145/2380116.2380139

Mambou, S., Krejcar, O., Maresova, P., Selamat, A., & Kuca, K. (2019). Novel hand gesture alert system. *Applied Sciences (Basel, Switzerland)*, *9*(16), 3419. doi:10.3390/app9163419

Najafabadi, M. M., Villanustre, F., Khoshgoftaar, T. M., Seliya, N., Wald, R., & Muharemagic, E. (2015). Deep learning applications and challenges in big data analytics. *Journal of Big Data*, *2*(1), 1–21. doi:10.118640537-014-0007-7

Nyirarugira, C., Choi, H. R., Kim, J., Hayes, M., & Kim, T. (2013, December). Modified levenshtein distance for real-time gesture recognition. In *2013 6th International Congress on Image and Signal Processing (CISP)* (Vol. 2, pp. 974-979). IEEE. 10.1109/CISP.2013.6745306

Ousmer, M., Vanderdonckt, J., & Buraga, S. (2019, June). An ontology for reasoning on body-based gestures. In *Proceedings of the ACM SIGCHI Symposium on Engineering Interactive Computing Systems* (pp. 1-6). 10.1145/3319499.3328238

Pandey, P., Dewangan, K. K., & Dewangan, D. K. (2018). Enhancing the quality of satellite images by preprocessing and contrast enhancement. *Proc. 2017 IEEE Int. Conf. Commun. Signal Process. ICCSP 2017,* 56–60. 10.1109/ICCSP.2017.8286525

Peng, Y., Tao, H., Li, W., Yuan, H., & Li, T. (2020). Dynamic gesture recognition based on feature fusion network and variant ConvLSTM. *IET Image Processing*, *14*(11), 2480–2486. doi:10.1049/iet-ipr.2019.1248

Perera, M., Haller, A., Méndez, S. J. R., & Adcock, M. (2020, November). HDGI: A Human Device Gesture Interaction Ontology for the Internet of Things. In *International Semantic Web Conference* (pp. 111-126). Springer. 10.1007/978-3-030-62466-8_8

Pham, C. (2015, October). MobiRAR: Real-time human activity recognition using mobile devices. In *2015 Seventh International Conference on Knowledge and Systems Engineering (KSE)* (pp. 144-149). IEEE. 10.1109/KSE.2015.43

Quek, F. (1994). Toward a Vision-Based Hand Gesture Interface. *Proceedings of Virtual Reality Software and Technology*.

Rahim, M. A., Islam, M. R., & Shin, J. (2019). Non-touch sign word recognition based on dynamic hand gesture using hybrid segmentation and CNN feature fusion. *Applied Sciences (Basel, Switzerland)*, *9*(18), 3790. doi:10.3390/app9183790

Sabour, S., Frosst, N., & Hinton, G. E. (2017). *Dynamic routing between capsules.* arXiv preprint arXiv:1710.09829.

Sahu, S. P., & Dewangan, D. K. (2021). Traffic Light Cycle Control using Deep Reinforcement Technique. *International Conference on Artificial Intelligence and Smart Systems (ICAIS)*, 697–702. 10.1109/ICAIS50930.2021.9395880

Schiphorst, T., Jaffe, N., & Lovell, R. (2005, April). Threads of recognition: Using touch as input with directionally conductive fabric. *Proceedings of the SIGCHI Conference on Human Factors in Computing Systems.*

Scoditti, A., Blanch, R., & Coutaz, J. (2011, February). A novel taxonomy for gestural interaction techniques based on accelerometers. In *Proceedings of the 16th international conference on Intelligent user interfaces* (pp. 63-72). 10.1145/1943403.1943414

Trigueiros, P., Ribeiro, F., & Reis, L. P. (2014). Vision-based Portuguese sign language recognition system. In *New Perspectives in Information Systems and Technologies* (Vol. 1, pp. 605–617). Springer.

Venkatnarayan, R. H., & Shahzad, M. (2018). Gesture recognition using ambient light. *Proceedings of the ACM on Interactive, Mobile, Wearable and Ubiquitous Technologies*, *2*(1), 1–28. doi:10.1145/3191772

Vincent, P., Larochelle, H., Lajoie, I., Bengio, Y., Manzagol, P. A., & Bottou, L. (2010). Stacked denoising autoencoders: Learning useful representations in a deep network with a local denoising criterion. *Journal of Machine Learning Research*, *11*(12).

Waibel, A., Hanazawa, T., Hinton, G., Shikano, K., & Lang, K. J. (1989). Phoneme recognition using time-delay neural networks. *IEEE Transactions on Acoustics, Speech, and Signal Processing*, *37*(3), 328–339. doi:10.1109/29.21701

Chapter 5

Recent Advancements in Design and Implementation of Automated Sign Language Recognition Systems

Bhavana Siddineni
SRM University-AP, India

Manikandan V. M.
SRM University-AP, India

ABSTRACT

Sign language has been used for a long time by deaf and mute people to communicate their thoughts and feelings. Since there is no universal sign language, the needy people use country-specific sign languages. For example, American Sign Language (ASL) is popularly used by Americans and Indian Sign Language (ISL) is commonly practised in India. Communication between two people who know the specific sign language is quite easy. But, if a mute person wants to communicate with another person who is not familiar with sign language, it is a difficult task, and a sign language interpreter is required to translate the signs. This issue motivated the computer scientist to work on automated sign language recognition systems that are capable of recognizing the signs from specific sign languages and converting them into text information or audio so that the common people can understand it easily. This chapter will be a useful reference for the researchers who are planning to start their research study in the domain of sign language recognition.

DOI: 10.4018/978-1-7998-9434-6.ch005

INTRODUCTION

Communication among people is the key to everything and language plays a crucial role in it. For effective communication, it is necessary to eradicate any language barriers. Most people are fortunate to be able to interact with everyone using words with sound. But some communities are unable to liaise similarly.

It is very difficult to imagine life without the ability to hear or speak. Anyone who cannot hear properly or people with hearing loss levels greater than 20 decibels are considered to be deaf. A person who cannot use their voice to speak is mute or dumb. Other than the deaf and mute sections there are people facing problems related to social interactions and repetitive behaviors, and with disorders like cerebral palsy, apraxia cannot communicate using generally spoken languages. It is also challenging for these communities to interact effectively with normal people.

All around the world, there are many such groups of people. The World Health Organization (WHO) estimates over 430 million people (5% of the total population in the world) (World Health Organization, 2021) suffering from deafness and hearing loss. Around 120 million people all over the world are both mute and deaf citizens. One in every ten individuals suffers from hearing and vocal impairments. The rate of the population of deaf-mutes caused by accidents, injuries, traumatic experiences, genetic disorders, and birth defects is rising rapidly. Hence it is important to address and answer the question how these sections of people face the world and communicate themselves.

Majority of people have all the amenities that a normal person must own. But the deaf and mute lack them and it is necessary to understand how they can communicate with the rest 90 percent of the population. It is unfair where one part of the world has the advantage of hearing whereas the other part is silent without any sound. It is heartbreaking and depressing when one cannot express and share their thoughts, feelings, ideas through words to their family members and friends and also when one cannot understand what others are communicating. It also limits their ability to perform daily tasks. This makes deaf-mutes feel stressed and frustrated physically, mentally, and emotionally. Most of them lose their self-confidence, become socially inactive, and isolate themselves.

Therefore, communication of deaf-mutes and other communities with ordinary people is mandatory. There are various corrective methods such as hearing aids, cochlear surgeries, etc. But all are not efficient and most of them are infeasible. These sections of people cannot communicate through spoken languages and need a distinctive way of communication with others. Other than the mouth and ears they can use their facial expressions and other body parts for conveying ideas. The most important and effective tool of communication among these groups deals with the use of sign language.

Sign language is also known as gesticulation language which is a mode of communication with people or animals through ocular processing by non-verbal means that involves movement of hands and limbs, the motion of the head along with the variations in frontal facial features and other body movements.

Sign language brought major transformation amidst deaf and mute people and it has the most intriguing history and evolution. The use of gestures was part of one of the oldest and primary ways of communication. Before the existence of formal sign language, hand movements were used for communication. During ancient times, three eminent philosophers Socrates, Plato, and Aristotle laid the foundation in writing the sign language employed by deaf and mute people. However, Aristotle believed that deaf-mute people cannot be educated. Contradicting this claim, Pedro Ponce de León in 16th century created formal sign language using monastery. Around the same time, Cardano's code gestures influenced many others. Later Bonet came up with proposals that were based upon demonstrative alphabets formed using the right hand and Aretina score. The first book on sign language released by Bonet fascinated everyone until the next revolutionary achievement. l'Eppe from France created hand signals to substitute the alphabets and found the first sign language school. In England, Thomas Braidwood established an institute for deaf and dumb and taught two-handed sign language. In 1690, people immigrated to Massachusetts from Kent County, England. Hereditary deafness became prevalent because of the unique genetics of these immigrants. Eventually, Martha's Vineyard turned out to be the place with the largest deaf population. On April 15, 1817, Thomas Hopkins Gallaudet established a school for deaf in America with his assistant Laurebt Clerc. American sign language was the ultimate result of developments in the signings of Gallaudet and Clerc. Next, a drastic decline in sign language was due to the decision of the Milan Conference in 1880. Sign language was banned and replaced with oral methods among the deaf. Despite it, there seemed to be growth in sign language, and further, The National Association of the Deaf came into existence. (Dayas, 2019; Seamons, 2017; Wikipedia Contributors, 2021a)

Sign language gained recognition as a method of communication for the deaf and dumb. Though some countries don't support sign language as an official language, it developed into a proficient language and made an impeccable impact on the lives of individuals. A common misconception of sign language being universal exists among people. The entire world doesn't use the same language for communication. Each continent, country, and states have their own set of languages and origins. Similarly, sign languages are influenced by people and their cultural backgrounds and so are diverse. Signs are created through the combination of facial grammar and hand orientation which are further organized specifically to convey the meaning of sentences. Nearly for every country on the globe, various sign languages prevail.

To increase awareness regarding the importance of sign language and support deaf and mute people United Nations General Assembly authoritatively declared 23rd September as the International Day of Sign Languages. Though there are 400 million deaf and dumb people all over the globe, it is surprising that the use of sign languages remains to be less than 10 percent of this population. Sign languages emerge naturally when groups of people communicate and share thoughts. The development of sign language depends on the geographical location and culture one follows. The dialects of sign language differ from country to country, region to region, and community to community. It is impossible to predict the total number of sign languages used worldwide. Sign language is not international and there is no accurate number butthere are around 300 different sign languages used by people. One sign language is not identical to another language. Every sign language is unique and independent with its own set of properties and grammar.

SIGN LANGUAGES AROUND THE WORLD

Based on region, there are many sign languages. All sign languages used in a location are not originated from the same area some of them are foreign sign languages. Africa has twenty-three sign languages at the minimum. There are around fifty different sign languages in use by people on the American continent. Asian and European regions also comprise nearly more than fifty sign languages each. The Middle East avail a variety of twenty-two different sign languages.

Sign languages are associated with families for classification. They are grouped based on the genetic correlations between languages. Their family trees include British, Danish, French, German, Japanese, and others. British tree is also known as Banzsl tree consists of three types British, Auslan, and New Zealand sign languages. The Swedish family is the branch of the British tree of sign languages. American sign language, Danish, Austro Hungarian, and Italian are parts of the French family tree (Wikipedia Contributors, 2021b).

British Sign Language is for the deaf communities in the United Kingdom. Old records reveal the existence of the British sign language in England around the 15th century. The establishment of Braidwood schools for the deaf and dumb took place in 1760. In 1974 BSL (British Sign language) was accepted. The language acquired recognition from the government in 2003. BSL has a particular grammar and syntax and is a spatial and structured visual language. One of the most significant components of BSL includes the use of pro-forms. It uses a peculiar and three sequenced concept comprising Time frame, topic, action, or comment. At first time frame is used to depict the variation of the words. Then a topic of discussion is established and is further expatiated in the comment section. Until it is marked again, the location in time and space is believed to remain the same. For education, many software and graphics of fingerspelling are being innovated by deaf tutors. For spelling scientific terms nearly 100 signs were appended to BSL in 2019. (Kyla, 2017; Neil & George, 2018)

The New Zealand sign languages of New Zealand and Auslan sign language of Australia evolved from BSL. They possess similar grammar and lexicons. The historical path of New Zealand sign language began in the 1880s. The NZSL Act of 2006, which gave people the right to use New Zealand sign language, was formally recognized by parliament. The foundational semantic unit of New Zealand Sign Language (NZSL) includes manual ones comprising signs formed by handshape, its location, the orientation of palms. Non-manual signs of the face to convey syntactic meaning were also utilized in NZSL (McKee, 2017). The two-handed signs, directional signs, classifier signs, compound signs along with lip-reading form the significant components of this language. The movements can be influenced by their direction, pace, style, and repetition. In contrast to English, NZSL tends to use particular rather than generic reference. With approximately 4000 signs, the Dictionary of New Zealand Sign Language made the language more accessible. British Sign Language and NZSL share 62.5 percent of their signs, compared to 33 percent of NZSL signs in American Sign Language (Wikipedia Contributors, 2021c).

Auslan came to Australia as a natural language from the variations of BSL by the immigrants and the teachers of both deaf and hearing. In 1991, Auslan was recognized as the language for the deaf community by the Australian government. Auslan has been influenced by the primary language English, particularly through fingerspelling and Signed English. BSL and Auslan are 79-90 percent related and identical. Other than Auslan there are various Australian sign languages like Walpiri and Yolungu.(Johnston & Schembri, 2001; Wikipedia Contributors, 2021d)

Charles Michel and de l'Epee devised a system of spelling words in French using manual alphabets. From this approach French sign language (old LSF) arose and is still used in France today. The Fabius statute, passed by the National Assembly in 1991, officially authorized the use of LSF in the teaching of deaf children. In 2005, a statute was passed that completely recognized LSF as a distinct language. The alphabet of French sign language is known as dactylogical and its grammar consists of three dimensions. At the same instant, one can express multiple ideas and concepts using FSL. The meaning of the phrase is revealed by the facial expression. FSL lacks conjugation but consists of time. Signer makes an action behind the shoulder for a past action, at the body level for a present, in front of the person for future action. The order of depiction involves time, place, and subject followed by the action. FSL is used by around 100,000 people as of 2019 and it has become the major pillar of deaf communities' identity. (Editors of Encyclopaedia Britannica, 2020; Wikipedia Contributors, 2021e)

Most people in and in regions around Denmark use Danish sign language. It was developed from the first deaf school of Denmark that was established by Castberg. Changed reference, shifted ascription of expressive parts, and shifted locus are three fundamental demonstrating ways of expressing a specific point of view in Danish Sign Language. In May 2008, the Danish Sign Language Dictionary was released consisting 2000 sign entries. Norwegian sign language is a hybrid of Danish sign language (DTS) and native sign. Icelandic sign language consists of only one-handed alphabets and is lexically 20 percent similar to Danish. (Engberg-Pedersen, 2008; Wikipedia Contributors, 2021f)

American sign language is the most widely used sign language in the world. It is genuine and precise and has many variations like other languages. History of sign language began with the establishment of American School for the deaf in Hartford. Signed English which was inspired from Signed French was followed. ASL has become the major language in deaf schools by 1835. Until Stokoe's arrival oralism persisted to be the method of deaf education educational approach involved bilingual bicultural model. ASL includes American manual alphabet with 26 gesture signs and 10 number signs that are used to spell large number of English words. The basic order of ASL is subject verb object language. It also allows sentences that do not follow the same order. Any aspect is indicated by changing the way the verb moves; punctual ones are indicated by a stationary hand position whereas continuous ones involve circular and rhythmic movement. ASL has a well-designed set of classifiers for objects and its movements. The dictionary of American Sign Language was published by William Stokoe. The number of ASL users in the United States is estimated to be between 100,000 and 15000,000 people (Drasgow, 2019; Wikipedia Contributors, 2021g).

The principal sign language among people in Netherlands is Dutch sign language (SLN or NGT), a descendent of old French sign language. In 1790, Henri Daniel institution, the first Dutch school was founded. The 1900-1980 years marked the prohibition period for sign usage. Later the use of signs rose with the establishment of Dutch schools and formation of dialects. NGT finally attained official recognition in October 2020.In Dutch sign language sentences begin with topic and the remaining part of the sentence is considered as comment. Along with the topic comment structure it possesses topic agreement instead of subject agreement. The first national Dutch sign language dictionary was published by Van Dale in 2009(Crasborn et al., 2009; Wikipedia Contributors, 2020).

A visual language prominent in Italy was Italian Sign Language (LIS). Most recently Italy recognized LIS on 19th May 2021. Signed Italian consists of the mixture of lexicons of LIS and spoken Italian. In LIS the concurrence between nouns verbs and adjectives is based on the location rather than gender. Nouns must be in the same order as pronouns and verbs and they can be inserted anywhere in the space.

For interrogative sentences, declaratives, imperatives and clauses facial expressions are depicted by the signers. In LIS, classifiers, present in nominal spheres, operate as pro-forms and influence the behavior of verbs. All over there are 40,000 Italian sign language speakers.(Chiara & Lara, 2020)

Sweden was possibly the first in the world to recognize bilingualism. The government acknowledged Sweden Sign Language in 1981. It is estimated that at least 10,000 people speak it as their primary language. Par Aron Borg, a Swedish educator and pioneer in the teaching of deaf people, initiated the development of Portuguese Sign Language (PSL) in 1823 at the Casa Pia in Lisbon. Borg founded the first deaf school in Portugal. Both Swedish and Portuguese sign languages are closely related. Other sign languages include Chinese, Japanese, Irish, German, Finish and African sign languages.

INDIAN SIGN LANGUAGE

India is one of the world's most populous countries. According to WHO, there are more than 630 lakhs of people in India who are deaf, with nearly 50 lakhs of them being children. Among them only 2.7 million people in India use Indian sign language (ISL) for communication. The country has only 700 sign language schools. There are nearly 300 certified sign language interpreters translating for a deaf population of millions (Sharma, 2018).

History of Indian Sign Language

The first reported use of ISL was in the nineteenth century, but it was associated with impairment far into the twentieth. In India before 18th century, signs were indicated by the gestures (also known as mudras) from the Natyasastra formed by the combination of facial expressions, hand and limb movements depicting various emotions and thoughts. However due to some beliefs deaf people were excluded from social situations. Debates on the views of deafness and sign language continued among various communities and as a result ISL rose to prominence.(Sinha, 2017)

A UK person named Dr. Heawe conceived the idea of school for the deaf in India. He sought the help of Lord Ripon, Viceroy of India for establishing a school for the deaf. Later a church archbishop of Bombay founded the first deaf school in Bombay in 1885. Deaf schools were started at Chennai and Kolkata in 1890 and at Mysore in 1902. Numerous schools continued to establish and all of them focused on oralism. Before partition in 1947, sign language was same across all regions. Later Indian sign language was borrowed from British sign language. It is also known as plains Indian sign language or Hand talk system. Deaf people in India after British imperialism started to speak for themselves and fought for their rights and their language culture (Wikipedia Contributors, 2021h).

Further UK teacher named T.A. Walsh arrived and gave a great impetus to education of the deaf. In 1977, a professor from Gallaudet University, USA named Dr Madan Vasishta, James Woodward and Susan De Santis started researching about Indian sign language by visiting deaf schools in the major cities of India. They observed signs used by deaf students in schools and found that there were some variations in the Indian sign language used in many states and published a book on ISL (The People's Linguistic survey of India, 2013).

In 1996, Dr Dilip Deshmukh after attending many deaf conferences, federations and lectures believed in bilingualism and its need in India and henceforth published "Sign language and Bilingualism in Deaf Education". In 2001, a national workshop for sign languages in Coimbatore was held where deaf signers from Maharashtra, Gujarat, Karnataka and other states were invited. The signs used by different deaf groups were observed and drew all of them from which the first Indian sign language dictionary was developed in 2001 with 1600 words from 42 cities and 21 states. With goal of providing deaf people the right to speech and expression and eradicate communicate barriers, the ministry for social justice and empowerment formally launched the ISL dictionary with 3000 terms. Most recently on 17th February 2021 the third edition with 10,000 words was released(New edition of Indian Sign Language dictionary to have 10,000 terms, 2012).

More on Indian Sign Language

It is necessary to use Indian Sign Language as the mode of deaf educationto facilitate communication of people with hearing impairment. Also, ordinary people must discuss languages like ISL and other scripts to avoid the deaf community's marginalization. Though Indian sign language possesses regional variations, a deaf individual belonging to one region can understand and reciprocate to the signs depicted by a person from another place.

The system used for coding language manually in India is known as Indian Sign System(ISS). It combines Indian Sign Language words (signs) with the order of words, syntax and grammar of at least six official Indian oral languages, including Signed Urdu, Hindi, Marathi, Telugu, and Tamil. In India, several types of sign language dialects are followed in different states. Mumbai sign language is the most popular and highly related lexically making it primarily suitable for compositions. The sign language dialects belonging to Chennai and Hyderabad are the next closely similar after Hyderabad and Mumbai variant. Calcutta's sign language differs greatly from that of Chennai.

Indian sign language is nonverbal, communication is done visually using a balance of manual and non-manual expression elements. Handshape, hand motion, and location are all manual characteristics. Whereas head and body position, gazing, lip movements and facial expressions are examples of non-manual characteristics. In India, research work has focused on determining the best attributes for correctly labeling a specific sign from a collection of all signs. In terms of ISL, the majority of the signs incorporate both manual and non-manual features.

- **Manual Features:** The four main criteria of Indian sign language manual features are hand shape, palm orientation, movement, and hand location. These parameters are used in numerous ways by various sign languages. As explained below, there are several movements used in the ISL.
  - **Finger Spelling:** A subtype of sign language that spells words with manual letters.
  - **Hand gestures:** They are a group of movements that use hands to communicate the messages. Handshape and movements are all used to define them.
  - **Face gestures:** They are a group of gestures that uses face to communicate the messages. This comprises the usage of brows, lip and facial expressions, and so on. Fixed gaze, expressions of surprise, delight, rage, grief, and so on are some examples.
  - **Body gestures:** They are a set of gestures that use the entire body to communicate information. Body expansion and body compression, are examples of these gestures.
- **Non-Manual Features:** Face expressions, gazing, lip movements, and body posture are some examples of non-manual features. A manual sign meaning can be modified using these features. The meaning of phrases is determined in part by facial expressions and body position. For example, the position of the eyebrows can indicate a question mark.

Every deaf individual must be aware of the signs of alphabets, numbers, weekdays, months, colors, relations and interrogative, and other words. Every sign is supposed to lie within the box region from head to chest and lie within the shoulder region. However very few exceptions exist. Also, it is a must to not speak while signing any word. These are the two major points to be kept in mind while learning and using sign language. Alphabets in every sign language have their own unique style and modifications. The most common way for people to begin learning sign language is to learn the A-Z or elements equivalent in sign form.

Grammar

The grammar of ISL is identical to that of other spoken languages. ISL's entire grammaticalization method is based on a layered approach. According to Zeshan (Zeshan, 2000), sign language has its own grammar standards. Later, Sinha (Sinha, 2017) emphasized numerous structures and ISL grammar usage. ISL grammar maintains the same norms throughout India, according to linguistic research, and each sign is either coinage or fingerspelling. Words, acronyms, and clips from a spoken language with no sign are introduced using fingerspelling. There are no inflections, conjunctions, verbs, articles in ISL. It only uses root words. In addition, to indicate the tense, it uses some unique spatial-temporal indices. The sign language used around the world differs. The grammar of Indian Sign Language is distinct from that of the manual representation of spoken English. It has different characteristics such as body gestures, mouth and hand gestures, eye movements, etc. The following are some of the features of ISL grammar:

Figure 1. ISL representations

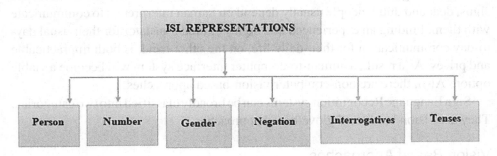

- **Person representation:** The first person is always addressed as the speaker, while the second person is always addressed as the addressee. Personal name signs are usually based on a person's physical characteristics (Sinha, 2017) as there are no pronouns in ISL.
- **Number representation:** The numbers are classified as single or plural. In ISL, the plural is accomplished in several ways. The most prevalent approach for pluralization is reduplication.
- **Gender representation:** Gender is conveyed in ISL by a symbol pointing to the side of the face (Sinha, 2017). It is articulated by a sign above lip level to communicate the sign for males. A female is represented with a fingertip upon on nose.

- **Negation representation:** In negative sentences, non-manual indications such as facial expressions and head motions are used to communicate negation.
- **Tenses representation:** Tenses are represented in ISL using signals, which can be used to illustrate all three tenses. The past tense, present tense, and future tense are represented by a spatial timeline (Zeshan, 2000). In ISL, the present tense is unmarked, but a spatial timeline depiction is employed to indicate the future or past tense of phrases.
- **Questions representation:** In addition to hand gestures, eye movements are used to communicate questions. For example, eyebrows when raised indicate that they are trying to convey questioning sentences.

Indian Sign Language Recognition Systems

Sign language automation, on the other hand, is much more than a collection of alphabets. The fact that these languages are multi-featured and multi-mode at the same time presents a challenge. Automation of Indian sign language is less significant than other languages as it comprises of two-handed signs. As the blind cannot see signs, deaf and dumb people usually depend on human interpreters to communicate with them. Finding an experienced and highly skilled translator for their usual day-to-day communication for their daily life, on the other hand, is both impracticable and pricey. As a result, a human-to-computer interface system will become a viable option. Also, there are non-computer vision-based approaches.

Sign Language Recognition systems can be broadly classified into two categories. They are vision based and glove-based approaches.

Vision Based Approaches

They are image or video-based interpretation models which are non-intrusive in nature. The main requirement of this method is the Camera. The common processes involved are preprocessing, feature extraction, classification, training and testing. Most of the vision-based approaches are based on human activity recognition. Both top down and bottom-up approaches of detecting key points for pose estimation with different types of neural networks and algorithms can be implemented to recognize and classify the signs displayed by any person(Banjarey et al., 2021).

Below flowchart depicts the use of numerous algorithms in vision-based approaches of Indian sign language recognition.

Figure 2. Vision-based approaches for ISL recognition

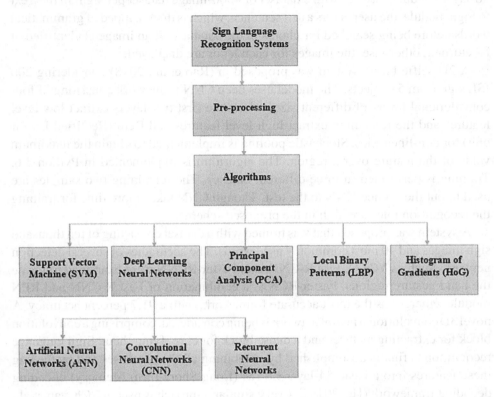

(Muthu Mariappan & Gomathi, 2019) used the Open-Source Library for Computer Vision to create a dynamic Indian sign language recognition system. Smoothing technique is used to preprocess resized frames, resulting in binary images with white areas representing skin regions Over them, other morphological alterations such as erosion and dilation are applied, followed by median blurring to remove noise. Face contours, right hand contours, and left-hand contours are discovered, and they form the regions of interest. The Fuzzy C means clustering algorithm is applied to the retrieved features, which divides them into clusters During testing, these clusters are compared to existing ones, and the gesture id with the highest degree is returned. For the study, data samples of 80 words and 50 sentences from everyday use were used.

In (Shrinidhi et al., 2020), a novel method was implemented which included two modules: sign to text (STT) and text to sign (TTS). The Sign to Text module consists of sub processes. Firstly, motion frame is identified using a background segmentation algorithm and then a frame detector is used to eliminate redundant frames. These are the two preprocessing steps. Further the pattern matching is performed using

50-layer residual network with a dataset of 4000-image dataset per sign. In the Text to Sign module, the user enters a text sentence, which is then stripped of grammatical words before being searched for photos in the database. If an image is identified, it is returned; otherwise, the images for each letter are displayed.

A 2D selfie-based system was proposed in (Rao et al., 2018) considering 200 ISL sign from 5 subjects. The model uses deep CNN framework consisting of four convolutional layers of different sizes where the first two layers extract low level features and the remaining extract high level features and Relu (Rectified Linear unit) for non-linearities. Stochastic pooling is implemented to obtain the maximum value of the feature over a region. The algorithm is implemented in Python 3.6. Training is performed in three different batches. The remaining two samples are used to put the trained CNN to the test. Though CNN takes more time for training the recognition rates are high in the proposed scheme.

A system was proposed that was trained with a dataset consisting of ten thousand sign images and 40 most commonly used words. The accuracies of several detection networks such as YOLO, Fast R-CNN, and Faster-RCNN were examined to locate the hand gesture regions. Faster-RCNN, a combination of Fast R-CNN and RPN module, emerges as the most accurate framework, with a 91.7 percent accuracy. A novel 3D convolutional neural network is being considered, comprising a convolution block for extracting features and a dense block for classifying them. Sign language recognition is finally accomplished by combining all the processes and by feeding these features into a basic 34-layer LSTM (Long Short-Term Memory) encoding decoding framework (He, 2019). A very similar approach is used in (Chavan et al., 2021). Additionally, a user interface was designed for the Indian Sign Language Interpreter using Adobe XD which captures, uploads, converts and gives an audio output.

With a similar 3D CNN network of filter size 25 x 224 x 224 as proposed in (Pardhi et al., 2021) for vehicles, one can even predict the movement of signs in sign language which will further aid in autonomous sign language recognition. Accurate results of visual movement can be obtained by implementing spatiotemporal organization. In (Dewangan et al., 2021) a VLDnet was proposed for intelligent vehicles, an implementation of similar approach consisting of the hybrid of UNet and ResNet can be designed to obtain an optimal sign region and sign detection. With this encoder decoder network of segmentation and sampling, we can identify the features and decode the segmented images for the dataset. A clear observation in any

A topographical descriptor was created (Kumar et al., 2020) using two color-coded pictures, joint distance, and joint angles obtained from joint locations in a study. For classification, a two coursed convolutional neural network that has been trained for 500 Motion - capture classes were considered. Both layers were integrated, and features from both areas were used. The database consists of various datasets that like HDM05. With less training time, the proposed methodology exhibits an increase in predictive performance. For sign language recognition, some researchers from (Kumar et al., 2020) and other authors (Kishore et al., 2018) proposed a technique that leverages motion lets. Segmentation divides a video frame into motion and non-motion joint sets, which characterize the features of the images obtained. The procedure is divided into three phases. The first phase creates the database by tracking hands and extracting their orientations, trajectories, and shapes. After that, for classification, an adaptive kernel technique is used. It resulted in a 98.9% recognition rate.

Another automated system of recognition suggested in (Kaur & Krishna, 2019) is simulated in MATLAB that uses SIFT for extraction and classification of features. Images are uploaded and database is created. Further the key points are obtained by the optimum of difference of Gaussians. The SIFT feature's distinctiveness is greater, but the possibility of a similar feature's presence exists. Using the fitness function of the ABC optimization method, the algorithm is utilized to select the preferable set of features. Further to train the proposed recognition system, selected feature sets are used as inputs to a Feed Forward Back Propagation Network (FFBPN) and accuracy of 99.43% is attained.

The Indian sign language alphabets recognition using deep learning was achieved in (Pardeshi et al., 2019). Patterns from images are directly identified using convolutional neural networks. Alex Net was one of the earliest architectures having eight layers. Google Net is a CNN with 22 layers and a parameter-minimization module. Both frameworks with average pooling strategy were implemented for the recognition of ISL alphabets in MATLAB using Deep Learning Toolbox. The training database consists of 5200 images and the parameters considered are size of the batch, learning rate, number of epochs and frequency of validation. Both had an accuracy rate of above 90%.

Sign language users usually don't get the chance to speak effectively during video conference sessions. The latest development from Google AI enables signers to speak and communicate and participate as active orators. Generally, real-time sign language detection is intensive computationally as it involves high-end heavy input tasks. Despite these obstacles, a paradigm for real-time recognition has been proposed. During the process of classification of each frame, the input dimensions are diminished by segregating every information that the system requires from the video to achieve a lightweight model. A pose estimation model is a computer vision-based approach that detects human movements from video and image inputs. With the help of this model, the input has reduced from a high-definition image to a compact set comprising of connected markings among the parts of hands, shoulders, nose, and eyes. A pattern of recognizable motion in every frame is computed using markings obtained and is further normalized. The system will recognize the signing if it falls within the threshold. The German Sign Language Corpus is used for testing. In addition, for predicting sign usage using optical flow, a linear regression model is trained. A deep learning recurring neural network architecture of long short-term memory consisting of memory of the preceding time series is used to classify and process data. When the user begins to sign then the person comes into speaker mode. When the model detects signing, an ultrasonic tone of voice of frequency less than 20Hz is sent via an audio cable.(Moryossef, 2020).

A mobile application named sign translator was developed in (Priya et al., 2020) that bridges the gap between regular people and those who are visually impaired. Its main back-end components are a speech recognizer, microphone, and image viewer. The system identifies the voice and turns it to a string, which is then presented alongside visuals that match the strinNears a result, normal individuals can express themselves and deaf and dumb persons can visualize them. It was created in Android Studio, with XML being used to create appealing interfaces and Java being used to implement the matching and recognition procedures.

A start-up developed an artificial intelligence and machine learning based application titled Gnosys which translates sign language to text or speech in real time. It is also known as the google translator and has provided cost effective plans to the users.

Analysts in (Sugandhi et al., n.d.) believe that trying to replace speech with sign language is a realistic solution. As existing datasets are insufficient for synthesizing signs directly from speech, an Indian sign language corpus of continuous speech encompassing 10,000 words of English vocabulary with 18 hours of recordings from 5 people was established first. The suggested model architecture is transformer-based, with continuous input and output vocabulary. It mainly consists of a joint speech encoder and pose and text decoders. Open Pose is used to generate poses, and back propagation is used to raise aspects to three dimensions. The speech encoder learns from the input speech and thereafter uses spatial encoding. The voice encoder and preceding steps are used to determine a pose at the current time step. The encoded input is fed into the text decoder, which decodes the tokens in the sequence. The correlations between speech and poses are identified using two modalities and a CM (cross modal) discriminator. A multi-task transformer network is also linked, which displays the sign sequences continuously for the specified voice sequence.

A set of 24 Indian sign language alphabets were recognized in (Singha & Das, 2013) that is based on the Euclidean distance and eigen values and vectors classification. The skin-coloured parts are extracted from the images through the filtering process and a blob image is generated as the result. Further the hands are cropped and the eigen values and eigen vectors are calculated from the covariance matrix of the input and are arranged with respect to the descending order of eigen values. Euclidean distance between the eigen vectors of the current and dataset images is computed and the minimum is considered. Next the product of this minimum and difference between eigen values is obtained. Results of all are summed and the minimum of all is the recognized symbol. An accuracy of 97% was observed in this method.

Another novel approach which uses discrete wavelet and discrete cosine transform for the recognition of Indian sign language. The features are extracted through DWT by image disintegration, decomposition of coefficients of low filter band and acquiring band coefficients. DCT can also be used to obtain features. The contribution of the domain component with high frequency is lessened and focuses on the energy over modified signal in low frequency. Further they are classified with the help of the dataset considered (Muthukumar, 2020). Same approach was followed along with Local Binary Patterns to increase the number of features. Local representation is generated by correlating every pixel with surrounded pixels (Muthukumar & Gobhinath, 2018).

A real time system based on grid features that ignores the face entirely and only recognises hand positions was developed in (Shenoy et al., 2018). For this study, a dataset of 24,464 sign pictures of integers from 0 to 9 and 23 letters was used. The model uses a K closest neighbour classifier to recognise poses, and the Hidden Markov Model to identify gestures. The initial stage in pre-processing is to extract hand features in each frame by removing the face, background, and skin segmentation using HOG and SVM. The slope of the line produced between the centroid of hands in consecutive frames determines the hand's motion. The feature extraction is carried out via grid-based fragmentation, and clusters for each hand posture are obtained. K nearest neighbours picks the class that occurs most frequently by calculating distance between samples. For gesture categorization, a discrete left to right Hidden Markov Model is utilized, which involves feeding encoded intermediate hand poses to model chains.

[46] aims to recognize single handed signs using histogram-oriented features and support vector machine classifier. For spell checking, the Natural Language Processing framework is employed. The proposed solution was tested using roughly 200 images and 10 sign videos in addition to the Kaggle datasets. For the training, a variety of machine learning models were used. In comparison to other models, HOG with SVM provides the highest level of accuracy. To generate a stronger matrix, canny edge detection is applied after the pre-processing Hog algorithm is used to count the number of times a gradient position appears in a particular area of an image and preserve the sign as a vector value. The normalising of the nine-bin vector computed from the horizontal and vertical gradients yields a Hog feature vector for a given sign image. Following all of these steps, the algorithm builds hyperplanes to divide the data into two groups. When the margin is at its maximum, the hyperplane is optimal. This technique also highlights and permits dual communication modes. Pyttsx3 is used to do a reverse recognition process. Pypspellchecker, an NLP module, is also

utilised to avoid misspellings and erroneous detections. A similar SVM algorithm is used for classification in (Raheja et al., 2016) that reduces computation speed by using Zernike moments for acquiring frames and an improvised technique for the removal of co-articulation to recognize single and double handed static and dynamic gestures along with the words of fingerspelling. In (Mali et al., 2019) Principal component analysis is used for the pre-processing stage. Principal component analysis reduces the dimension by considering the feature possessing largest variance. An SVM classifier is used for the classification.

It is highly important to at least recognize signs of words used in emergency situations. One automated recognition approach considers hand gestures of eight different words taken from 26 people. The eight emergency words include accident, call, doctor, help, hot, lose, pain and thief. Both features based (Multiclass SVM) and neural network based (LSTM with Google Net) models were implemented for these words and an accuracy of 90% and 96% was achieved in both the cases. (Adithya & Rajesh, 2020)

Kinect generates the skeleton of human body and recognizes joints in it. Microsoft Kinect is most used because performance is not affected by other external and surrounding conditions. In (Raghuveera et al., 2020) a dataset with 4600 images with 140 gestures belonging to ISL from 21 people captured using Kinect box was taken into account. Kinect sensor is used to separate the hand region from the depth image. After pre-processing using K means clustering, SVM is used by assembling with three other models and the output is finally predicted.

Table 1 depicts the summary of all the Vision Based approaches discussed above along with the accuracy rates.

Table 1. Summary of vision-based approaches

S.No	Method In	Description	Dataset	Accuracy
1.	(Muthu Mariappan & Gomathi, 2019)	Open source, fuzzy C means	80 words, 50 sentences	75%
2.	(Shrinidhi et al., 2020)	Res Net	4000 images per sign	99.09%
3.	(Rao et al., 2018)	Deep CNN, RELU with Stochastic pooling	200 ISL signs from 5 subjects	92.88%
4.	(He, 2019)	Fast RCNN and RPN module	10000 sign images and 40 words	99%
5.	(Kumar et al., 2020)	topographical descriptor joint locations and two coursed CNN	HDM05 500 motion capture sets	95%

Continued on following page

Table 1. Continued

S.No	Method In	Description	Dataset	Accuracy
6.	(Kishore et al., 2018)	motion and non-motion joints adaptive motion let kernels	500-word dataset	98.9%
8.	(Kaur & Krishna, 2019)	sift and feed forward back propagation	ISL alphabets and numbers from 130 subjects	99.43%
9.	(Pardeshi et al., 2019)	Deep Learning with CNN, MATLAB	5200 images	Alex net – 98.61% Google net – 91.9%
10.	(Moryossef, 2020)	optical flow feed through LSTM linear regression, pose estimation model	Sign language corpus	91.5%
11.	(Priya et al., 2020)	Mobile App, Sign Translator, Speech Recognizer, Microphone and image viewer	Strings and image viewer	
12.	(Sugandhi et al., n.d.)	Open Pose, text speech voice encoders and decoders, CM discriminator, Multi Task transformer network	18 hrs recordings from 5 people	95%
13.	(Singha & Das, 2013)	Euclidean distance and eigen values and vectors classification	24 ISL signs	97%
14.	(Muthukumar & Gobhinath, 2018)	DCT and DWT with Local Binary Patterns	ISL alphabets, signs, numbers and gestures	87%
15.	(Shenoy et al., 2018)	Hidden Markov Model, K closest neighbour.	33 hand poses and 12 gestures from ISL	Static-99.7% Gestures – 97.23%
16.	(Manjushree & Divyashree, 2019)	NLP modules, HOG with SVM, Canny Edge Detection	200 images and 10 sign videos	97.1%
17.	(Raheja et al., 2016)	Zernike moments and SVM	80 signs, real time capturing video	97.5%
18.	(Mali et al., 2019)	PCA for pre-processing SVM for classification		95.52%
19.	(Adithya & Rajesh, 2020)	Multi class SVM and LSTM	8 emergency words	SVM-90% LSTM 96%
20.	(Raghuveera et al., 2020)	Microsoft Kinect, K Means, SVM with three feature classifiers	4600 images, 140 gestures from 21 subjects	71.85%

GLOVE BASED APPROACHES OF ISL

Till now we have discussed vision-based approaches for sign language recognition that were non-intrusive and cost effective. Now we move on to another set of approaches based on glove systems. The glove based are intrusive and require large number of hardware components like sensors and detectors. However, glove-based approaches have a significant advantage over vision-based systems in regard to the direct sending of voltage values and data which reduces the need of processing the raw data.

The components used for development of system in (Bhat, 2015) are flex sensors, Microcontroller AT89S52, Liquid Crystal Display, Analog to Digital Converter, Bluetooth, Smartphone. To trace the hand movement the gloves are built with flexible sensors and a supply of voltage is passed. Due to the bending of the sensors a voltage fluctuation is observed. The obtained analogy voltage is transformed to digital voltage and then is sent to microcontroller. The microcontroller validates and displays the output text of the sign on the LCD screen by correlating with pre-determined voltage structures. The textual characters acquired are further sent to a mobile application designed using MIT app inventor through Bluetooth for conversion of the text to speech.

One advantage of glove-based approaches is that it is not hindered by external conditions like light, background, etc. In (Bairagi, n.d.) along with the resistance sensors accelerometers are also used. The apparatus consists of ADXL335 accelerometer, Bluetooth module HC-05, Android phone and 10 variable sensors. The resistance sensors are placed on the rear side of the fingers along with accelerometer sensors mounted on the back part of the hands. A text to speech android application with options of menu connect, disconnect and clear is developed. A microcontroller LPC2138 takes the output from both sensors and generates ASCII values one by one. After acquiring values for all ten, the system as a whole check with information form accelerometer sensor to classify it as a meaningful gesture and a database is also created. Finally voice signal is given as output through mobile.

A translating glove named talking hands was proposed in (Heera et al., 2017) where along with flex and accelerometer additionally a gyroscope is utilized to obtain accurate results. The angular movement of the hands in space is measured using the gyroscope and the orientation across three different axes is determined by the accelerometer. An MPU6050 is used for the purpose. Instead of the microcontrollers Arduino boards are used for the development since they are more cost effective and supports all operating systems. A similar procedure from above mentioned approaches is followed. Software application is supported with SQLite database of signs and words of Indian sign language. A similar approach is presented in (Rewari et al., 2018) with an additional SD card which maps data and enables the addition of huge storage.

Both vision and glove-based methodologies include human activity recognition. While most of the deep learning approaches outlined in the preceding section functioned well, managing time-series data remains a difficulty. Improvement in pre-processing and feature extraction can be observed and implemented in sign language recognition through wearable sensor data and further classification of it with the combination of bidirectional long short-term memory and convolutional neural networks. This can aid in the development of high-performance models (Challa et al., 2021). Without the involvement of any feature extraction techniques and minimal pre-processing and utilizing the sign activity data obtained from gyroscopes or accelerometers automated feature extraction and classification can be achieved through convolutional nueral network and gated recurrent units as followed in (Dua et al., 2021)

INDIAN DATASETS

A sign language dataset is a collection of data and information in the form of sign images and videos which is used to analyse, evaluate, train and test any model. As already discussed, sign languages in India differs from region to region. There is no standard sign language in India. Due to which there is no common dataset which the signers can follow. Below mentioned are some of the Indian sign language datasets.

- **IITA-ROBITA:** In July 2009, the artificial intelligence and robotics lab of Indian Institute of Technology, Allahabad, developed this database of ISL signs for gesture identification with the moto of promoting advancements in the field of sign language recognition systems. Gestures are captured using Sony handycam at 30 fps with a consistent background. The resolution of gesture images is 320 x 240 pixels.
- **INCLUDE:** Recently in October 2020 an Indian Lexicon Sign Language Dataset – INCLUDE was proposed. It consists of 2,70,000 frames spanning over 4,292 videos with 263-word signs from 15-word categories. The videos were taken from the signers of Louis Deaf School, Chennai, ensuring near similarity to natural circumstances. Each recording depicts a single sign. Both training and testing datasets are available. The training set consists of 3475 files while the testing one contains 817 videos. By choosing a set of optimal hyper-parameters a batch of 50 words were considered and formed a separate dataset titled INCLUDE-50. It was proposed with the aim of quick assessment of the sign language recognition methods. A total of 958 videos

were acquired where the training dataset of include-50 has around 766 videos and the test set of 192 videos. As a part of study in (Sridhar et al., 2020) these datasets were implemented on various deep learning models and identified an accuracy rate of 85.6%for the include dataset and 94.5% for the include-50 subset the high performing model.

- **ISL-CSLTR:** Indian Sign Language Dataset for Continuous Sign Language Translation and Recognition was released recently in January 2021. For the advancements in sign language translation and recognition (SLTR), a thoroughly named sentence staged dataset was created. It was made with the help of two native signers from Andhra Pradesh's Nava Jeevan Residential School for the Deaf and amateurs from Sastra University. For 100 Spoken language Sentences performed by 7 distinct Signers, the ISL-CSLTR corpus contains a vast vocabulary of 700 fully labelled recordings, 18863 Sentence images, and 1036-word sign frames. Canon digital single lens reflex camera was used to record the videos. This corpus is organized based on signers and temporal constraints, and it is made publicly accessible including fully annotated details. This corpus was created to meet the many issues that SLRT researchers confront and increases interpretation and detection performance dramatically

A dataset for depth images was proposed in a report. There were 43,750 images with depth in all, with 1,250 for each of the 35 hand motions. These were taken from five people. The movements encompass all alphabets and digits. The photos are monochromatic and 320 x 240 pixels in resolution. Another Indian sign Language dataset is in Kaggle consisting of all digits from 0-9 and letters A-Z in 35 directories with 1200 files. It also suggests that the optimal strategy is to train using a big dataset, such as the ImageNet Large Scale Visual Recognition Challenge (ILSVRC), and then tune it for ISL later.

An Indian sign language corpus resource is presented by the organization, Linguistic Data Consortium for Indian Languages (LDC - IL). A total raw data of length 20 hours is available as the dataset. Every recorded video is captured from three different angles. The corpus is classified into five groups. A 6hrs 30 min data of free conversation, 3hrs 50 min of jokes, 5 hrs 34 min of stories, 4 hrs 15 min on sentences and sign words.

There are other datasets available for various sign languages across the world. Based on the type of sign language some of them are discussed below.

The American sign language lexical video set (ASSLVD) includes the data obtained from University of Boston from six native signers depicting around 9800 tokens. Another motion capture dataset of animations, CUNY ASL was generated in 2014. It helps the deaf with least English skills. A MSR gesture 3D dataset captured with

Kinect tracker consists of 12 variable signs subject to ten people. All in all, there are 337 files. A Purdue ASL database was created which was arranged in three sub databases and is accessible only through HD'S or DVD's. A BSL corpus consisting of 50,000 signs obtained from four different regions and a BSL sign bank was formed. German data sets include DGS Kinect 40 with 3000 samples, RWTH PHOENIX WEATHER with 45760 samples, SIGNUM with 33210 samples. Chinese sign language data sets are DEVISIGN -G, DEVISIGN-D, DEVISIGN-L containing 36, 500 and 2000 classes of rgb videos respectively. PSL 101 is a polish sign language dataset. There are many more kinds of datasets present all over the world.

CHALLENGES

This section discusses the various challenges faced in the recognition of Indian sign language. One of major drawback for the study in this area is the inexistence of a standard dataset for Indian Sign language alphabets, Mumbai and Pune being so close to each other have different sign languages. The formation and development of a standard dataset makes the model more resilient to sign changes and would undoubtedly enhance the findings. Also, majority of the Indian signs are two handed which make the process of recognition more difficult. Some pairs of alphabets like 'v', '2' and 'w' and '3' share the same sign which makes the differentiation hard. (e, f), (m, n), (u,v,w) are closely related gestures that are frequently misinterpreted. Moreover, few signs in Indian Sign language have motion associated with them making the recognition process more complex. The position, orientation and way of expressing signs vary from person to person which affects the pre-processing stage of any sign language recognition system. A negative influence is exerted by the presence of background variations and lighting in real time recognition or an improper dataset. One can overcome the drawback of coloured images and illuminations by semantic segmentation using fully convolutional neural network, U-Net and Seg-net to distinguish sign and non-sign regions in an image. A similar technique was followed in (Reddy et al., 2021). The efficiency of the system sometimes reduces as majority of the signs possess overlapping structural representations. The optimal approach for acquiring images is still being researched. Researchers also encounter potential limitations while using classification algorithms. Researchers are unable to concentrate on one optimum strategy due to the large number of recognition methods available. Selecting one methodology to focus on results in the testing of alternative methods that may be more suited for Sign Language Recognition. Researchers who experiment with multiple ways rarely develop one model to its highest capacity. A significant setback of glove-based approaches is that it requires the signers to wear them all time which is cumbersome due to attachment of many sensors and these sensors are not economical. Also, the sizes of hands vary from every person and may give false results if undersized or oversized gloves were used by the signers. Hence vision-based approaches are more preferred. Most research works considered only the manual features. Hence it is highly required to develop systems that identify and detect the non-manual features along manual features of the signer to achieve complete sign language recognition. The main objective and essentiality lie in the practical implementation of these models as they provide assistance in public places like malls, restaurants, etc. Another drawback includes the leakage of any secret data. Techniques of cryptography must be implemented in such situations. There is no proper awareness among the signers regarding the latest recognition systems and technologies available.

CONCLUSION

This chapter mainly focuses on automated sign language Recognition Systems on Indian sign languages. Communication of deaf and mute people among themselves and others is necessary. Importance of sign language and the evolution of sign language is highlighted. Further we went on describing the various types of sign languages followed all over the world comprising British, Australian, Indian, American, Italian, Swedish, New Zealand and others. A complete description about Indian Sign Language including its importance, history, features and grammar have been discussed. This paper presents the various vision and glove-based sign language recognition approaches developed in the recent times. The vision methodologies based on machine learning, artificial intelligence and computer vision were elucidated. The models include the implementation of Support vector machine, Histogram of gradients, Deep learning techniques applying convolutional neural networks, Hidden Markov Model, and wavelet transforms and other software applications. The glove-based techniques consisting the use of different sensors like flex, accelerometers, gyroscopes discovered by various analysts were reviewed. Further we extended and explained about the various sign language datasets like ROBITA, INCLUDE, etc. Also, the challenges faced for the research, development and innovation of sign language recognition systems in the future were acknowledged. The chapter deals with the literature review of different recent recognition systems with the end goal of providing researchers, analysts, beginners and interested individuals in the development of sign language systems a useful reference and a wide perspective in this area.

REFERENCES

Adithya, V., & Rajesh, R. (2020). Hand gestures for emergency situations: A video dataset based on words from Indian sign language. *Data in Brief, 31*, 106016. doi:10.1016/j.dib.2020.106016 PMID:32715044

Bairagi. (n.d.). Gloves based Hand Gesture Recognition Using Indian Sign Language. *International Journal of Latest Trends in Engineering and Technology, 8*(4), 131-137. doi:10.21172/1.841.23

Banjarey, K., Prakash Sahu, S., & Kumar Dewangan, D. (2021). A Survey on Human Activity Recognition using Sensors and Deep Learning Methods. *2021 5th International Conference on Computing Methodologies and Communication (ICCMC)*, 1610-1617. 10.1109/ICCMC51019.2021.9418255

Bhat. (2015, May). Translating Indian sign language to text and voice messages using flex sensors. *International Journal of Advanced Research in Computer and Communication Engineering, 4*(5). DOI doi:10.17148/IJARCCE.2015.4593

Challa, S. K., Kumar, A., & Semwal, V. B. (2021). A multibranch CNN-BiLSTM model for human activity recognition using wearable sensor data. *The Visual Computer*. Advance online publication. doi:10.100700371-021-02283-3

Chavan, A., Ghorpade-Aher, J., Bhat, A., Raj, A., & Mishra, S. (2021). Indian sign language interpreter for deaf and mute people. *International Journal of Creative Research Thoughts, 9*(3), 1389-1393. Available at https://www.ijcrt.org/papers/IJCRT2103178.pdf

Chiara, B & Lara, M. (2020). *A Grammar of Italian Sign Language (LIS)*. Fondazione Università Ca' Foscari. doi:10.30687/978-88-6969-474-5

Crasborn, O., van der Kooij, E., Ros, J., & de Hoop, H. (2009). Topic agreement in NGT (Sign Language of the Netherlands). *Linguistic Review, 26*(2-3), 355–370. doi:10.1515/tlir.2009.013

Dayas. (2019, May 28). How monks help create sign language. *National Geographic*. https://www.nationalgeographic.com/history/history-magazine/article/creation-of-sign-language

Dewangan, D. K., & Sahu, S. P. (2021). Road Detection Using Semantic Segmentation-Based Convolutional Neural Network for Intelligent Vehicle System. In K. A. Reddy, B. R. Devi, B. George, & K. S. Raju (Eds.), *Data Engineering and Communication Technology. Lecture Notes on Data Engineering and Communications Technologies* (Vol. 63). Springer. doi:10.1007/978-981-16-0081-4_63

Dewangan, D. K., Sahu, S. P., Sairam, B., & Agrawal, A. (2021). VLDNet: Vision-based lane region detection network for intelligent vehicle system using semantic segmentation. *Computing*, *103*(12), 2867–2892. Advance online publication. doi:10.100700607-021-00974-2

Drasgow, E. (2019). American Sign Language. In *Encyclopedia Britannica*. https://www.britannica.com/topic/American-Sign-Language

Dua, N., Singh, S. N., & Semwal, V. B. (2021). Multi-input CNN-GRU based human activity recognition using wearable sensors. *Computing*, *103*(7), 1461–1478. doi:10.100700607-021-00928-8

Editors of Encyclopaedia Britannica. (2020, November 12). Sign language. In *Encyclopedia Britannica*. https://www.britannica.com/topic/sign-language

Engberg-Pedersen, E. (2008). Point of View in Danish Sign Language. *Nordic Journal of Linguistics*, *15*(2), 201–211. doi:10.1017/S0332586500002602

He, S. (2019). Research of a Sign Language Translation System Based on Deep Learning. *2019 International Conference on Artificial Intelligence and Advanced Manufacturing (AIAM)*, 392-396. 10.1109/AIAM48774.2019.00083

Heera, S. Y., Murthy, M. K., Sravanti, V. S., & Salvi, S. (2017). Talking hands — An Indian sign language to speech translating gloves. *2017 International Conference on Innovative Mechanisms for Industry Applications (ICIMIA)*, 746-751. 10.1109/ICIMIA.2017.7975564

Johnston, T., & Schembri, A. (2001). *Australian Sign Language (Auslan): An Introduction to sign language linguistics*. Cambridge University Press.

Kaur & Krishna. (2019, August). An Efficient Indian Sign Language Recognition System using Sift Descriptor. *International Journal of Engineering and Advanced Technology, 8*(6). doi:10.35940/ijeat.F8124.088619

Kishore, P. V. V., Kumar, D., Sastry, A. S. C. S., & Kumar, E. K. (2018). Motionlets Matching With Adaptive Kernels for 3-D Indian Sign Language Recognition. *IEEE Sensors Journal*, *18*(8), 3327–3337. doi:10.1109/JSEN.2018.2810449

Kumar, E., Kishore, P. V. V., Kiran Kumar, M. T., & Kumar, D. A. (2020). 3D sign language recognition with joint distance and angular coded color topographical descriptor on a 2 –stream CNN. *Neurocomputing*, *372*, 40–54. doi:10.1016/j.neucom.2019.09.059

Kyla, A. (2017, April 23). *A Simple Introduction to BSL Grammar*. appa.me.uk. https://appa.me.uk/a-simple-intro-to-bsl-grammar/jyh

Mali, D., Limkar, N., & Mali, S. (2019). Indian Sign Language Recognition using SVM Classifier. *Proceedings of International Conference on Communication and Information Processing (ICCIP)*. https://ssrn.com/abstract=3421567 doi:10.2139/ssrn.3421567

Manjushree, K., & Divyashree, B A. (2019, July). Gesture Recognition in Indian Sign Language using HOG and SVM. *International Research Journal of Engineering and Technology, 6*(7).

McKee. (2017, April). Assessing the Vitality of New Zealand Studies. *Sign Language Studies, 17*, 322-362. https://www.researchgate.net/publication/316920727_Assessing_the_Vitality_of_New_Zealand_Sign_Language doi:10.1353/sls.2017.0008

Moryossef, A. (2020, October 1). *Developing Real-Time, Automatic Sign Language Detection for video conferencing*. Google AI blog. https://ai.googleblog.com/2020/10/developing-real-time-automatic-sign.html

Muthu Mariappan, H., & Gomathi, V. (2019). Real-Time Recognition of Indian Sign Language. *2019 International Conference on Computational Intelligence in Data Science (ICCIDS)*, 1-6. 10.1109/ICCIDS.2019.8862125

Muthukumar. (2020, December 1). Performance based algorithm for DWT and DCT for ISL. *Materials Today: Proceedings*. 10.1016/j.matpr.2020.10.639

Muthukumar, P. S., & Gobhinath, S. (2018, April). Extraction of Hand Gesture Features for Indian Sign languages using Combined DWT-DCT and Local Binary Pattern. *International Journal of Engineering and Technology*. doi:10.14419/ijet.v7i2.24.12072

Neil, P., & George, S. (2018). British Sign Language: Facts and Information. *Disabled World*. https://www.disabled-world.com/disability/types/hearing/communication/british-sign-language.php

New edition of Indian Sign Language dictionary to have 10,000 terms. (2012, February 17). *The Hindu*. Retrieved from https://www.thehindu.com/news/national/new-edition-of-indian-sign-language-dictionary-to-have-10000-terms/article33855541.ece

Pardeshi, K., Sreemathy, R., & Velapure, A. (2019). Recognition of Indian Sign Language Alphabets for Hearing and Speech Impaired People using Deep Learning. *Proceedings of International Conference on Communication and Information Processing (ICCIP)*. https://ssrn.com/abstract=3430055 doi:10.2139/ssrn.3430055

Pardhi, P., Yadav, K., Shrivastav, S., Sahu, S. P., & Kumar Dewangan, D. (2021). Vehicle Motion Prediction for Autonomous Navigation system Using 3 Dimensional Convolutional Neural Network. *2021 5th International Conference on Computing Methodologies and Communication (ICCMC)*, 1322-1329. 10.1109/ICCMC51019.2021.9418449

Priya, L., Sathya, A., & Raja, S. K. S. (2020). Indian and English Language to Sign Language Translator- an Automated Portable Two Way Communicator for Bridging Normal and Deprived Ones. *2020 International Conference on Power, Energy, Control and Transmission Systems (ICPECTS)*, 1-6. 10.1109/ICPECTS49113.2020.9336983

Raghuveera, T., Deepthi, R., Mangalashri, R., & Akshaya, R. (2020, January 30). A depth-based Indian Sign Language recognition using Microsoft Kinect. *Indian Academy of Sciences.*, *1234*(1), 34. Advance online publication. doi:10.100712046-019-1250-6

Raheja, J. L., Mishra, A., & Chaudhary, A. (2016). Indian Sign Language Recognitionusing SVM. *Pattern Recognition and Image Analysis*, 26(2), 434–441. doi:10.1134/S1054661816020164

Rao, G. A., Syamala, K., Kishore, P. V. V., & Sastry, A. S. C. S. (2018). Deep convolutional neural networks for sign language recognition. *2018 Conference on Signal Processing And Communication Engineering Systems (SPACES)*, 194-197. 10.1109/SPACES.2018.8316344

Rewari, H., Dixit, V., & Batra, D. (2018). Automated Sign Language Interpreter. *Proceedings of 2018 Eleventh International Conference on Contemporary Computing (IC3)*.

Seamons, S. (2017, April 19). *The History of Sign Language*. https://aslblog.goreact.com/the-history-of-sign-language/

Sharma, A. (2018, September 25). Breach the wall of silence: Give State recognition to Indian Sign Language. *Hindustani Times*. Retrieved from https://www.hindustantimes.com/analysis/breach-the-wall-of-silence-give-state-recognition-to-indian-sign-language/story-hg7lj7LTWzfKgYB19prOhP.html

Shenoy, K., Dastane, T., Rao, V., & Vyavaharkar, D. (2018). Real-time Indian Sign Language (ISL) Recognition. *2018 9th International Conference on Computing, Communication and Networking Technologies (ICCCNT)*, 1-9. 10.1109/ICCCNT.2018.8493808

Shrinidhi, G., Amina, B., Hasan, H., & Ramsha, A. (2020, July). Indian Sign Language Translator Using Residual Network and Computer Vision. *International Research Journal of Engineering and Technology, 7*(7).

Singha, J., & Das, K. (2013). *Indian Sign Language Recognition Using Eigen Value Weighted Euclidean Distance Based Classification Technique*. ArXiv, abs/1303.0634.

Sinha, S. (2017). *Indian Sign Language: An Analysis of Its Grammar*. Gallaudet University Press.

Sridhar, A., Ganesan, R. G., Kumar, P., & Khapra, M. M. (2020). INCLUDE: A Large Scale Dataset for Indian Sign Language Recognition. *Proceedings of the 28th ACM International Conference on Multimedia*. 10.1145/3394171.3413528

Sugandhi, Parteek, & Sanmeet. (n.d.). Sign Language Generation System Based on Indian Sign Language Grammar. *ACM Transactions on Asian and Low-Resource Language Information Processing, 19*. doi:10.1145/3384202

The People's Linguistic survey of India. (2013). *Sign Language*. Orient Blackswan. https://lms.jamdeaf.org.jm/pluginfile.php/1489/mod_resource/content/1/The_Peoples_Linguistic_Survey_of_India_S.pdf

Wikipedia Contributors. (2020, December 23). Dutch Sign Language. In *Wikipedia, The Free Encyclopedia*. Retrieved 03:58, July 14, 2021, from https://en.wikipedia.org/w/index.php?title=Dutch_Sign_Language&oldid=995990988

Wikipedia Contributors. (2021a, May 11). History of sign language. In *Wikipedia, The Free Encyclopedia*. Retrieved 12:06, June 20, 2021, from https://en.wikipedia.org/w/index.php?title=History_of_sign_language&oldid=1022589456

Wikipedia Contributors. (2021b, June 28). List of sign languages. In *Wikipedia, The Free Encyclopedia*. Retrieved 12:11, June 21, 2021, from https://en.wikipedia.org/w/index.php?title=List_of_sign_languages&oldid=1030852077

Wikipedia Contributors. (2021c, January 7). New Zealand Sign Language. In *Wikipedia, The Free Encyclopedia*. Retrieved 08:08, June 21, 2021, from https://en.wikipedia.org/w/index.php?title=New_Zealand_Sign_Language&oldid=998864012

Wikipedia Contributors. (2021d, March 25). Auslan. In *Wikipedia, The Free Encyclopedia*. Retrieved 05:25, June 21, 2021, from https://en.wikipedia.org/w/index.php?title=Auslan&oldid=1014236843

Wikipedia Contributors. (2021e, July 12). French Sign Language. In *Wikipedia, The Free Encyclopedia*. Retrieved 07:45, July 13, 2021, from https://en.wikipedia.org/w/index.php?title=French_Sign_Language&oldid=1033242667

Wikipedia Contributors. (2021f, June 30). Danish Sign Language. In *Wikipedia, The Free Encyclopedia*. Retrieved 16:30, July 13, 2021, from https://en.wikipedia. org/w/index.php?title=Danish_Sign_Language&oldid=1031269623

Wikipedia Contributors. (2021g, July 12). American Sign Language. In *Wikipedia, The Free Encyclopedia*. Retrieved 03:01, July 14, 2021, from https://en.wikipedia. org/w/index.php?title=American_Sign_Language&oldid=1033241599

Wikipedia Contributors. (2021h, July 7). Indo-Pakistani Sign Language. In *Wikipedia, The Free Encyclopedia*. Retrieved 06:28, July 15, 2021, from https://en.wikipedia. org/w/index.php?title=Indo-Pakistani_Sign_Language&oldid=1032460482

World Health Organization. (2021, April 1). *Deafness and Hearing Loss*. https:// www.who.int/news-room/fact-sheets/detail/deafness-and-hearing-loss

Zeshan, U. (2000). *Sign Language in Indo-Pakistan: A Description of a Signed Language*. John Benjamins Publishing Company. doi:10.1075/z.101

Chapter 6
Hidden Markov Model for Gesture Recognition

Aradhana Kumari Singh
School of Computer Science, University of Petroleum and Energy Studies, India

ABSTRACT

In this chapter, hidden Markov model (HMM) is used to apply for gesture recognition. The study comprises the design, implementation, and experimentation of a system for making gestures, training HMMs, and identifying gestures with HMMs in order to better understand the behaviour of hidden Markov models (HMMs). One person extends his flattened, vertical palm toward the other, as though to reassure the other that his hands are safe. The other individual smiles and responds in kind. This wave gesture has been associated with friendship from childhood. Human motions can be thought of as a pattern recognition challenge. A computer can deduce the sender's message and reply appropriately if it can detect and recognise a set of gestures. By creating and implementing a gesture detection system based on the semi-continuous hidden Markov model, this chapter aims to bridge the visual communication gap between computers and humans.

INTRODUCTION

The HMM is dependent on the Markov chain being supplemented. A Markov chain is a model that explains the probability of sequences of random variables, or states, each of which can take on values from a range of possibilities. Words, tags, or symbols representing anything, such as the weather, can be used to create these sets. A Markov chain makes the extremely strong premise that if we wish to forecast the future of a sequence, we must first predict the past. In similar way HMM can be

DOI: 10.4018/978-1-7998-9434-6.ch006

used in Human Computer Interaction in many way, following sub-section motivates us for with many ways.

Motivation

One person extends his flattened, vertical palm toward the other, as though to reassure the other that his hands are safe. The other individual smiles and responds in kind. This wave gesture has been associated with friendship from childhood. Human motions can be thought of as a pattern recognition challenge. A human expresses motion patterns in order to communicate visual messages to a receiver. These patterns, which are loosely referred to as gestures, are variable but distinct and have a meaning. Because even the same person's hand position can differ by several inches from that of a prior wave, the wave gesture is changeable. It stands out because it can easily be separated from other gestures like beckoning or shrugging. Finally, it has the universally agreed-upon connotation of "hello." There are a variety of reasons for studying gesture recognition, many of which are related to increasing inter-personal communication.

A computer can deduce the sender's message and reply appropriately if it can detect and recognise a set of gestures. A conductor, for example, can direct a "virtual orchestra" by waving directions to a video camera. As a result, the system adjusts the volume and pace of the prerecorded music being played.

By creating and implementing a gesture detection system based on the semicontinuous Hidden Markov Model, this chapter aims to assist bridge the visual communication gap between computers and humans. The HMM framework models the temporal behaviour of gesture, and the use of a global codebook allows the discovery of shared atomic pieces among different gestures, leading to a "understanding" of gesture by identifying their similarities and differences.

BACKGROUND

To understand terminology HMM used for gesture recognition purpose we first need to understand the problem encounter in this area of research. Now here, we can divide the problem into two major areas: recognition of pattern and gestures.

The part on pattern recognition, for example, provides a solid framework for the HMM gesture identification problem by describing the pattern recognition approach, data clustering, and the Hidden Markov Model. Following that, a description of gesture recognition wraps up the background information and goes through some related studies in the field.

Pattern Recognition

Pattern recognition forms the mathematical basis of gesture recognition in this chapter. First, a definition of pattern recognition establishes the necessary terminology. Next, clustering, a process for grouping similar objects together, is described. Finally, the foundation of this chapter, the Hidden Markov Model is presented.

Definition

Pattern recognition is a precise mathematical field that categorises items into one of several groups. Print characters, voice waveforms, textures, system "states," and anything else that can be classified are all examples of patterns. The pattern recognition method is usually set up in such a way that it can recognise patterns without the need for human intervention. For instance, a system could alert a credit card firm to transactions that are most likely the result of illegal credit card use. Learning from a set of sample patterns is required to build a pattern recognition system. There are two types of learning processes. The learning process is referred to as supervised pattern recognition if the classes of the example patterns are already known. In this scenario, the system's performance is judged based on the proper classification of a single pattern. This input enables the system to develop itself iteratively. If the classes aren't known ahead of time, the unsupervised pattern recognition system must not only generate a classification algorithm, but also define the classes. Despite the fact that pattern recognition of this type is substantially more challenging, helpful algorithms have been discovered that allow successful systems to be constructed. The pattern recognition process is divided into two phases, the first of which is feature extraction, in which a pattern's observation is changed into a vector, the components of which are known as features. is often considerably more tractable for the system than, but it should contain the majority of the information required for pattern categorization. The approaches for feature extraction may be merely mathematical strategies for simply lowering the dimensionality of the observations, or they may be based on intuition or physical considerations of the problem. The feature vectors are classified in the second phase of pattern recognition. A classifier divides the feature space into discrete sections, each of which corresponds to a different pattern class (Duda & Hart, 1973). If a specific observation's feature vector falls within the region, the observation is classified. As a result, the partition determines the observations' class membership. Because the class membership of a set of example patterns is known, constructing the classifier in a supervised pattern recognition system is quite simple. As previously stated, this information is utilised to train and test the classifier. Construction is substantially more complicated and includes clustering if class membership is unknown, the same is describe in next section.

Figure 1. The pattern recognition process.

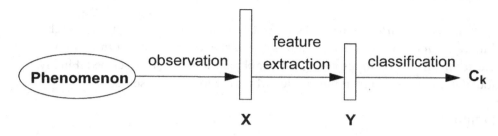

The pattern recognition procedure is depicted in Figure 1. First, while seeing a phenomenon, an observation vector is recorded. The significant aspects of the original observation are encapsulated in a smaller vector created through feature extraction. Lastly, categorization determines the most appropriate class label for. A video sequence, for example, or a sequence of hand parameters or another measured property, or a class of waving gestures. A system can determine the phenomenon's class in this way. To put it another way, the system is aware of the occurrence.

Clustering

Clustering is the process of creating an unsupervised pattern recognition classifier. The difficulty is not only to classify the given data, but also to define the classes at the same time. Clustering is the method of creating an unsupervised pattern recognition classifier. The difficulty is not only to classify the given data, but also to define the classes at the same time. Clusters are defined as groups of comparable points based on some measure of similarity in the broadest sense. Typically, similarity is defined as the proximity of points in the feature space as assessed by a distance function, such as the Euclidean distance. Measures of other qualities, such as vector direction, can, however, be utilised as well. The approach for locating clusters may be based on a heuristic or on the minimization of a mathematical clustering criterion. Vector quantization is clustering utilising the Euclidean distance metric in the field of digital signal processing; nevertheless, several additional words are utilised. The quantization levels of a VQ code book are now known as a classifier's clusters. Furthermore, the distance between each sample and the mean of its enclosing cluster is now a measure of distortion rather than similarity. Vector quantization aims to discover the collection of quantization levels that minimises average distortion across all samples. Finding the codebook with the least average distortion, on the other hand, is impossible. Nonetheless, given the number of clusters K, the simple K-means algorithm can achieve convergence to a local minimum:

1. Assign samples to clusters at random.
2. Calculate each cluster's sample mean.
3. Assign each sample to the cluster with the most similar mean.
4. If all of the samples' classifications remain the same, quit. Otherwise, proceed to Step 2.

Figure 2. Data Clustering. (a) data collection, (b) The clusters and their associated mean are visualised as a vector quantization., and (c) Gaussian distribution estimation

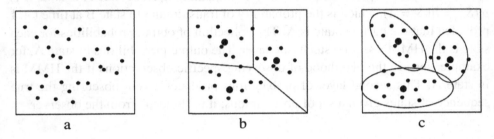

a   b   c

Hidden Markov Models

Table 1.

Codebook	
cluster	pdf
C_1	0 1
C_2	0 1

The Markov process, which is the cornerstone of the Hidden Markov Model, must first be described. There is usually enough structure in any pattern to influence the

113

probability of the next event. In the English language, for example, the likelihood of detecting the letter u is highly dependent on whether the letter q was previously identified, because u almost always follows q. If the conditional probability density of the current event, given all past and present occurrences, depends solely on the j most recent events, the stochastic process is called a jth-order Markov process.

A Hidden Markov Model (abbreviated as λ) is a process that is doubly stochastic. The underlying first-order Markov process, as shown in Figure 2.3a state transition diagram, is the first stochastic layer. Each state represents a possible observation of the Markov process, and the probability of transitioning from state A to state B is $p\,(S_{t+1} = B|\, S_t = A)$, which is the probability of transitioning to state B at time t + 1 provided that the current state is A. The collection of output probabilities for each state is the HMM's second stochastic layer. The output probabilities of state A, for example, specify the likelihood of observing specific observations if the HMM is in state A. This second layer of probabilities produces a veil, obscuring the true sequence of states given a set of observations; it is "hidden" from the observer.

Figure 3. Markov models. (a) First-order Markov process, (b) Discrete Hidden Markov Model (HMM), (c) Continuous HMM, and (d) Semi-continuous HMM.

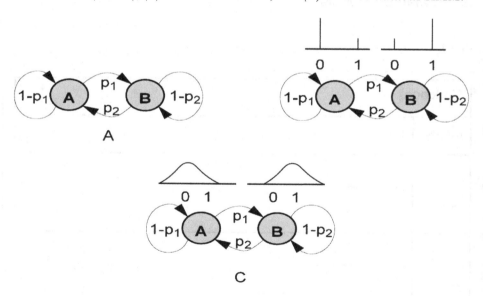

There are algorithms for both training and testing HMMs. The goal of HMM training is to lift the veil so that the real sequence of states S can be inferred with a high degree of certainty from the series of observations X. However, sufficient training data must be provided in order to construct a good internal statistical model.

The Baum-Welch re-estimation approach, which has been shown to converge, can identify locally optimal HMM parameters for a given set of training data. Initial estimates that are reasonable might aid the method in finding the global best option. The Viterbi algorithm considers the state sequence S with the highest probability given a certain observation sequence X (i.e., it maximises P S X(λ)) for testing and recognition.

Three fundamental forms of HMMs have been identified in the literature, each distinguished by the approach used to describe output probabilities. The discrete HMM's observations are discrete symbols from a finite alphabet that typically correlate to a vector quantization codebook's quantization levels (classes). Each state has a discrete probability mass function (PMF) that describes the likelihood of the state producing a certain symbol. Figure 2.3b depicts an HMM with two output symbols: state A is more likely to create a "0" symbol than a "1," whereas state B is more likely to generate a "1" symbol than a "0." The likelihood of viewing particular multidimensional, continuous data is represented by a mixture of probability density functions (pdfs) in each state of the continuous HMM. To accurately characterise the state's membership in the space of observation vectors, Gaussian (normal) pdf mixtures are commonly utilised. 2.3c illustrates an HMM with a single 1-D pdf for each state; state A is still more likely to yield a "0" than a "1," but the continuous 1-D observations are no longer distorted by quantization. A semicontinuous HMM is a cross between a discrete and a continuous HMM. The observation vectors are quantized into one of a finite set of classes, similar to the discrete HMM, minimising the number of free parameters. The observation classes, like the continuous HMM, are modelled by a multivariate Gaussian pdf, which removes the quantization distortion and effectively models the variance of an observation class. This formulation is analogous to a continuous HMM with parameter tying, in which states must share the same pdfs. The classes are reestimated along with the HMM parameters to build an integrated model, which is initially formed by clustering the example data. Figure 2.3d displays a two-state HMM with a two-cluster code book; while the observations are still described by 1-D pdfs, all states share these pdfs. The semicontinuous Hidden Markov Model is used in this study. The classes are reestimated along with the HMM parameters to build an integrated model, which is initially formed by clustering the example data. Figure 2.3d displays a two-state HMM with a two-cluster code book; while the observations are still described by 1-D pdfs, all states share these pdfs.

Gesture Recognition

This section provides the final background information. The concepts of gesture and gesture recognition are first defined. Following that, previous work on gesture recognition is summarised.

Definition

The system's definition of gesture then becomes a probabilistic definition that relies on higher-level knowledge from outside the system to define gesture. Explicitly, gestures are specified as sets of trajectories through a quantifiable gesture space within the system. The above definition is based on a number of significant assumptions. To begin, the measuring feature space reflects the complete space of gestures. As a result, the expressiveness of gesture space is heavily dependent on the features utilised; if the features are improper for the task at hand, or if there are too few features, the system will be unable to discern between distinct motions. Gestures, on the other hand, are modal in nature. Modality indicates that the collection of trajectories for each gesture is mutually exclusive because a gesture instance is a trajectory across feature space. This assumption is also based on the features that have been selected. Finally, gestures can be statistically characterised as a series of perceptual shifts (states defined by feature measurements). Perceptual states are feature space clusters of observation vectors. This assumption arises from the usage of HMMs, and it is more of an issue for producing than recognising examples of gestures. When determining the type of gesture an HMM represents, for example, a modest transition probability may provide a prototype that is completely different from the typical example.

Literature on HMM

Many research studies on learning and recognising visual behaviour have been published. However, given to its newness in the vision community, only a tiny number of Hidden Markov Models have been published. Below is a list of some of the more interesting works relating to this topic.

The majority of early vision research involving HMMs was limited to handwriting recognition, as seen in (Nag, et al., 1986). (Starner et al., 1994), employed an established HMM voice recognition product for a real-time handwriting recognition system recently.

(Yamato et al., 1992), employed discrete HMMs to successfully distinguish six different tennis swings among three people in work relating to human gesture. Despite the fact that the system only used 25x25-pixel images as the feature vector, it was able to discern between the various motions. However, the domain had to be limited enough to prevent the effects of vector quantization distortion in order to employ the discrete HMM. Furthermore, the system could only recognise isolated gestures from video that had already been temporally divided.

Sign language, which has a well-defined vocabulary and syntax, is probably the most clear example of human gesture. (Starner & Pentland, 1995), used HMM approaches from the speech community to recognise American Sign Language. Starner insisted that the signer sit in a specific position, wear coloured gloves, and refrain from signing with his fingers. These constraints resulted in a single 8-dimensional feature vector containing the x and y positions of each hand, the angle of the axis of least inertia, and the eccentricity of the enclosing ellipse. He also used strong grammar to get 97 percent accuracy with this small feature vector.

"A self-organizing framework... for learning and detecting spatiotemporal events (or patterns) from intensity image sequences,". (Cui et al., 1995), defined their system. They used this approach to recognise a single hand's hand sign. Despite the fact that they carefully differentiated between the most expressive and discriminating traits, all of them were essentially eigenvector coefficients. The sample gestures were used to divide the visual space into partitions, each with its own set of eigenvectors. The gesture was believed to be isolated inside a temporal window, from which five frames were sampled to depict the gesture. They claimed a 96 percent recognition rate for a simple vocabulary, but acknowledged to tinkering with a threshold to get there. Despite the fact that it is not built on HMMs, it is useful to show how much complexity HMMs contain by looking at a system comparable to this theory that does not.

(Bregler & Omohundro, 1995), provided a technique for lip reading problems that was an interesting application of HMMs. Their study was significant since it included auditory and visual information in an HMM system. Their solution was view-based to keep the vision task easy. Smooth nonlinear manifolds were used to model the human lip from the photos. Using the additional visual information, they observed considerable gains in continuous speech recognition in noisy conditions.

A gesture, according to (Bobick & Wilson, 1995), is a series of states in a measurement space. Unlike HMMs, this model allowed a prototype trajectory to be calculated to reflect the gesture. By explicitly defining two categories of variation, along-trajectory variance and across-trajectory variance, this prototype was utilised to align the state pdfs along the most-likely direction. This approach is significantly different from HMMs because the prototype allowed tracking of the gesture within a state, whereas HMMs cannot model the motion within a state.

(Wilson & Bobick, 1995) combined numerous representations for each state using a state-membership metric. The Baum-Welch algorithm was used to reestimate the HMM parameters, which was interspersed with reestimation of the representation parameters. They were given more latitude in defining the representations, such as the use of eigenimages, by employing the measure of state membership instead of the conventional pdf's in HMMs. They state, however, that this approach may prevent convergence of the reestimated HMM parameters, even if this has not been seen.

Hidden Markov Models for Gesture Recognition

The HMM recognizer application creates a new code book of a defined size and a set of HMMs (one per gesture) with a specified number of states after the training data is generated and saved as a test suite. The HMMs are produced with an option of left-right or ergodic transition matrix, and the code book is clustered on the observation vectors using a modified K-means algorithm.

The HMMs are immediately tested on the training data when the system parameters have converged to ensure that the training is accurate. The recognizer can also be checked with previously unseen gesture samples at this point. When an effective recognizer is developed, it may be simply applied to samples of unknown type, which is the ultimate goal of pattern recognition.

Summary

Feature extraction lowers raw observation vectors to feature vectors, clustering divides the feature space into perceptual states, and Hidden Markov models divide the feature space into perceptual states. The temporal behaviour between those states is represented by models. The definition of a gesture influences gesture recognition. As a result, system performance is strongly reliant on the quality of the features retrieved, yet gesture recognition can be reduced to a purely computational form. The domain of 2-D mouse motions was used to design and test a method for training and testing semicontinuous HMMs.

HMMs have a number of intriguing and relevant behaviours that have been discovered through preliminary testing. First, we're reminded that the code book's initial random grouping encourages us to think about multiple training sessions before drawing conclusions. Second, we can observe that the system's performance is highly dependent on the quality of the features picked. Absolute location and velocity were enough for recognition in the easy studies; however, the slightly more difficult 5-letter recognition challenge could not be successfully applied to very identical independent test suites. Furthermore, the fundamental contrast between representative and distinctive characteristics necessitates a rethinking of the HMM formulation's state output modelling. We've also seen that, while an HMM system may have learned to recognise the training set, it may not have learned any underlying, intuitive qualities for reliably classifying new observations.

CONCLUSION

Accurate recognition does not guarantee accurate modelling in the statistically-based Hidden Markov Model. The trained HMM identifies the training set and related testing sets despite the fact that it does not make intuitive sense. However, these HMMs fail catastrophically on slightly different test sets because to a critical balance that fails with previously discovered variances, not only recognising many ludicrous patterns as well as actual patterns. Many systems rely on large training sets to enumerate the variations in order to maintain this important balance; yet, collecting and categorising all of the data is expensive. (Huang, et al., 1990) semi-continuous HMM makes the similarities between patterns obvious, which was originally driven by minimising the amount of free HMM parameters. This technique offers two major advantages over traditional continuous HMMs. First, accurate modelling results in a robust, variance-insensitive HMM that does not require a significant quantity of data. Second, the global codebook highlights the commonalities between gestures, allowing for a better "knowledge" of gesture relationships. This study examined the behaviour of HMM training and revealed some key findings based on multiple experiments with 2-dimensional mouse motions.

The (Baum, 1972), (Baum et al., 1970) re-estimation process has more latitude to choose an erroneous model because mixtures of Gaussians provide a better approximation of a control state's observation membership. Figure 4.1 depicts three distinct-behaving gesture segments that during training could be mistaken for a blend of Gaussians. We discovered that supplying enough states for training, followed by state merging to limit the HMM to a canonical form, could prevent this problem in a left-right HMM. Other HMM architectures should be given more study in order to avoid this problem. Experiments with smarter data selection approaches to reveal the minimum data set for a successful recognizer would be welcome. Second, when the complexity of gestures increases, checking the values of system parameters becomes cumbersome.

They, discovered a few crucial principles on how to use HMMs. To begin with, because training is an expensive technique, we feel that every effort should be made to limit the amount of training data required to get acceptable results. This can be aided by better features and the removal of comparable cases. Finally, the quality of the features can sometimes be the sole determinant of a recognition system's performance. We propose a strategy for dividing up the feature space into discrete feature subspaces rather than forcing the system to use undesirable features. This can be done by hardcoding zeros in the code book's covariance matrices to force statistical independence, or by constructing a new code book for each feature subspace. The latter strategy appears to the author to be the most promising.

Figure 4. Inequivalent HMM structures cause confusion. All three structures can be misconstrued as a single state with a combination of Gaussians, despite their highly diverse behaviour. (a) indicates a relationship that is parallel (exclusive-or), (b) displays a juxtaposition (sequential) relationship and (c) The relationship is nondeterministic (equivalence). Only for (c) is a model like this accurate.

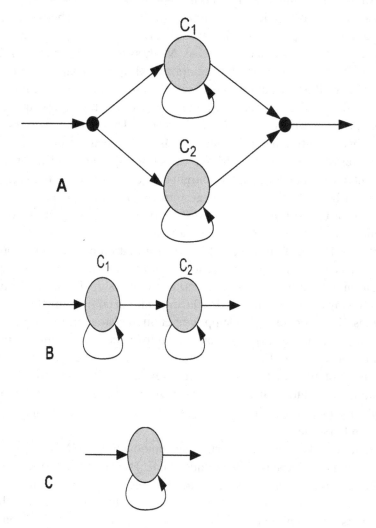

HMMs have promise, but they lack a few key characteristics. First, they have no concept of a prototype, which prevents gesture summarization. Rather to calling an HMM by its given name, it would be much more useful to call it by a common gesture. Second, we notice a trade-off in the studies here between identifying representative characteristics, which leads to an understanding of gesture similarities, and discovering distinguishing features, which leads to gesture distinction. More research into this issue is needed, as well as suggestions for how to improve the HMM. Finally, because state transitions are probabilistic, large errors such as omission can go undiscovered.

An "a" is, for example, a "d" with a short stem, which causes the "d" HMM to score well just by crossing a state transition quickly. Introduce the concept of state duration, as explained in (Rabiner & Juang, 1986), then this type of difficulty may be avoided.

FUTURE WORK

Many improvements and extensions have been obvious during the course of the study, as with any research effort. To begin, more research in the realm of 2-D mouse gestures should be conducted. Most significantly, more robust and usable data such as angle, change in angle, and relative position of the mouse rather than absolute position should be collected. This will increase the results of recognition. Furthermore, a fascinating addition to the system would be the ability to separate feature subspaces. Keep the position and velocity features distinct, for example, to reduce the number of system parameters (by deleting entries from the code book's covariance matrices) and possibly improve convergence. Of course, claiming that two subspaces of the feature space are orthogonal will have significant ramifications, necessitating further investigation. Create two independent code books, one for each feature subspace, to expand on this concept. Finally, additional typical modifications to the HMM architecture, such as time length modelling and corrective training, may aid performance.

Execution performance can be enhanced in two ways, regardless of processor speed. The first way is to directly optimise the C++ programme and generate optimised code using compiler settings. A second approach is to make the problem simpler. Large feature vectors, for example, may be substituted with smaller ones, or observation sequences could be undersampled to reduce the number of observation vectors by half. Using the discrete HMM instead of the semicontinuous HMM is another option to simplify the problem. The discrete HMM reestimation equations are more simpler than the semicontinuous HMM reestimation equations. Comparing the trade-offs between discrete HMM simplicity and semicontinuous HMM expressiveness would be a fascinating research.

The creation of tools for visualisation of HMM methods is a more practical expansion with an immediate requirement. Visualization, regardless of the area in which HMMs are used, would aid in a better understanding of the system dynamics. For instance, if an experimenter observes that the HMM parameters have already converged, the training process can be halted. The parameter matrices can be shown in this way by showing them as images, with the brightness of each image position corresponding to the relative weight of each entry in the matrix. When a person sees not just the strongest HMM for describing an example, but also its strength in comparison to other HMMs, the label assigned to the example in recognition can be qualified. Visualization can also help you understand the code book. When a two-dimensional code book is utilised, the 2-D pdfs in the feature space are displayed in a window. A mouse gesture feature space with two codebooks, for example, can be represented in two windows, one with a 2-D subspace for location and the other with a 2-D subspace for velocity. When 2-D features aren't present, just presenting the covariance matrices can help determine whether an independent feature subspace exists, allowing for the creation of a distinct code book.

REFERENCES

Baum, L. (1972). An inequality and associated maximization technique in statistical estimation of probabilistic functions of markov processes. *Inequalities, 3*, 1–8.

Baum, L., Petrie, T., Soules, G., & Weiss, N. (1970). A maximization technique occurring in thestatistical analysis of probabilistic functions of markov chains. *Annals of Mathematical Statistics, 41*(1), 164171. doi:10.1214/aoms/1177697196

Bobick, A., & Wilson, A. (1995). Using configuration states for the representation and recog-nition of gesture. In *Proc. Fifth International Conf. on Computer Vision.* IEEE Press.

Bregler, C., & Omohundro, S. (1995). Nonlinear manifold learning for visual speech recogni-tion. In *Proc. Fifth International Conf. on Computer Vision.* IEEE Press. 10.1109/ICCV.1995.466899

Cui, Y., Swets, D., & Weng, J. (1995). Learning-based hand sign recognition using SHOSH-LIF-M. In *Proc. Fifth International Conf. on Computer Vision.* IEEE Press.

Duda, R., & Hart, P. (1973). *Pattern Classification and Scene Analysis.* John Wiley & Sons, Inc.

Huang, X., Ariki, Y., & Jack, M. (1990). *Hidden Markov Models for Speech Recognition.* Edinburgh University Press.

Nag, R., Wong, K., & Fallside, F. (1986). Script recognition using hidden Markov models. ICASSP 86. doi:10.1109/ICASSP.1986.1168951

Rabiner, L., & Juang, B. (1986). An introduction to hidden markov models. *IEEE ASSP Magazine*, *3*(January), 4–16. doi:10.1109/MASSP.1986.1165342

Starner, T., Makhoul, J., Schwartz, R., & Chou, G. (1994). On-line cursive hand-writing recognition using speech recognition methods. ICASSP 94.

Starner, T., & Pentland, A. (1995). Visual recognition of American Sign Language using hiddenmarkov models. *Proc. of the Intl. Workshop on Automatic Face- and Gesture-Recognition.*

Wilson, A., & Bobick, A. (1995). Learning Visual Behavior for Gesture Analysis. *Proc. IEEE Symposium on Computer Vision.*

Yamato, J., Ohya, J., & Ishii, K. (1992). Recognizing human action in time-sequential images usinghidden markov models. In *Proc. 1992 IEEE Conf. on Computer Vision and Pattern Recognition*. IEEE Press. 10.1109/CVPR.1992.223161

Chapter 7
A Six–Stream CNN Fusion–Based Human Activity Recognition on RGBD Data:
A Novel Framework

Kamal Kant Verma
Uttarakhand Technical University, India

Brij Mohan Singh
College of Engineering, Roorkee, India

ABSTRACT

RGBD-based activity recognition is quite an interesting task in computer vision. Inspired by the exemplary results obtained from automatic features learning from RGBD data, in this work a six-stream CNN fusion approach has been addressed, which is developed on 2D-convolution neural network (2DCNN) and spatial-temporal 3D-convolution neural networks (ST3DCNN). The proposed approach has six streams and runs in parallel, where the first and second streams are used to extract space and time features with the help of a ST3DCNN model. Similarly, the remaining four streams have been used to extract the temporal features by means of two motion templates on motion history image (MHI) and motion energy image (MEI) via a 2DCNN. Further, a support vector machine (SVM) is employed to generate the score from each stream. Finally, a decision level fusion scheme particularly a weighted product model (WPM) to fuse the scores is obtained from all the streams. The effectiveness of the proposed approach has been tested on popular benchmark public datasets, namely UTD-MHAD, and gives promising results.

DOI: 10.4018/978-1-7998-9434-6.ch007

INTRODUCTION

Automatic recognition of human activities is dynamic research problem in machine vision field from past two decades. The objective of HAR is to develop a system to recognize the activity performed by the human body parts either in indoor or outdoor scenes. The human activities may vary from a basic level to the more complex one. For an instance, activities like eating, drinking, jogging or clapping are the simple one while the activities such as watching TV, cooking and read the book are the more complex activities. An activity is represented as the continuous variation in the appearance and motion sequences. The proposed system recognizes these activities performed in the video sequences or in the stack of images. HAR system has wide range of uses in various domains like intelligent monitoring systems, health care systems, pedestrian monitoring systems, abnormal behavior analysis, robotics and human machine interaction etc. Furthermore, it is a difficult task to recognize the class of each activity regardless to the changes in subject's structure and the similarity between the activity classes. Apart from the aforementioned difficulties, additional external conditions such as camera motion, multi-camera resolution, occlusion, human-object interaction, and human-human interactions make the video-based activity recognition system more difficult. People can gain a better understanding of human behavior by systematically analyzing and recognizing the human postures. This could be useful for augmenting the process of observing the indoor activities, understanding unusual activities, and maintaining a real-life security surveillance system. As a result, human activity recognition is an interesting problem in today's scenarios.

In the recent years, there has been a substantial enhancement in this field due to the two major reasons: first, the frequent access to the depth cameras, and second, the automatic learning of features via the deep CNN. Before the development of the RGBD camera, the RGB video cameras were the only ones capable of detecting human activities (Aggarwal & Ryoo, 2011). As low-cost RGBD cameras such as the Microsoft Kinect, Xbox 360 Kinect, and ASUS Xtion became available, thus the research has shifted to the learning of human activities using depth cameras (Xia et al., 2012). The existences of low-cost depth camera also resolve the problem of HAR at the greater extent and provide the cost-effective solution. In addition to the availability of depth camera, the wide accessibility of multi-model and heterogeneous datasets such as UTKinectAction3D (Xia et al., 2012), MSRAction3D (Li et al., 2010), UTD-MHAD (Chen et al., 2015), CAD-60 (Sung et al., 2012) and NTU-RGBD (Shahroudy et al., 2016) datasets have considerably achieved the success in the field of HAR, using RGB-D data.

In computer vision, a deep neural network (DNN) has achieved a considerable improvement due to its deep architecture over the past few years. A DNN is a systematic organization of several layers that automatically extract features from input raw data. In DNN, a layer obtains the input from the output of the previous layer and uses a non-linear transformation. (CNN) is one such type of deep neural network that achieves better recognition results. CNN achieved sufficient popularity in vision-based applications during the computer vision era. Normally, CNN is divided into two types: 2D-Convolutional Neural Network (2DCNN) and spatio-temporal 3D-Convolutional Neural Network (ST3DCNN).

Figure 1. Flow diagram of the proposed approach

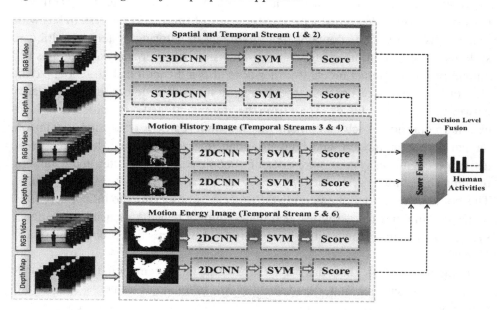

The 2DCNN learns visual features from 2D-images while ST3DCNN is capable to learn both visual appearance and motion features form the video inputs simultaneously. From the past few years ST3DCNN achieved considerable success in human activity recognition system (Ji et al., 2012). ST3DCNN is enough capable to learn 3D human behavior using RGB-D data (Zhao et al., 2019). However, the combination of spatio-temporal feature learning by CNN and classification using support vector machine (SVM) has extraordinary advantage. CNN-SVM has successfully implemented for various applications such as handwritten digit classification (Niu & Suen, 2012), cancer detection (Xue et al., 2016), and human activity recognition (Sargano et al., 2017).

Recently, researchers have utilized motion history and motion energy images from the RGB videos (Bulbul et al., 2019; Khaire et al., 2018; Ma et al., 2018), gait and gestures recognitions (Sarma et al., 2020). These images contain the prior knowledge and keep regularly updating the most recent information. Due to this, they can comfortably deal with moving subjects such as human activity, action and gestures. Both MHI and MEI do the motion analysis task in videos. MHI retains the meaningful motion information by generating the silhouette from the grayscale images. Therefore, it can efficiently represent the motion sequences. MHIs are less affective to noises of silhouette for example shadows, variable illuminations and holes. A MHI assign the brightest intensity value to the pixel that is just moved and lowest intensity describe the value to the pixel that has not moved for the long time denoting the previous motion. Basically, MHI contains the entire history of motion sequence within the video into a single template where intensity value of each pixel changes over time. Motion energy images (MEIs) are the reformation of MHIs. MEIs describes the entire motion in a single block, irrespective of the direction of movement. Both MHI and MEI represent the overall clue of the motion sequence without considering the segmentation of the body part movement. MHI and MEI generate less overhead during computation making them more efficient and useful for real time activity processing (Sarma et al., 2020). Inspired by the above facts, to find the motion present in the videos and the direction of the motion, we have adopted MHI and MEI motion template matching for activity recognition from the RGBD data.

In previous studies, mostly the activity recognition from RGBD data has been performed using two separated streams, one stream to learn the visual features from the RGB videos and in the second stream the temporal saliency features have been obtained from depth maps. Therefore, this chapter proposes a six-stream approach in which visual and temporal features learning have been performed from both RGB videos and depth map together. Instead of learning spatial and temporal features separately, we used ST3DCNN to learn both visual and temporal features simultaneously from RGBD data in the first and second streams. The remaining four temporal video streams have also utilized with the help of MHI and MEI obtained from the RGBD using 2DCNN, and this is the most distinguishable segment of our suggested approach and should get enough attention.

Likewise, deep multi-stream fusion approaches (Cong & Zhang, 2020) perform much better as compare to the traditional methods (Khoshelham & Elberink, 2012). Thus, the idea behind this work is to suggest multi-stream fusion approach which extracts both spatial and temporal features from both RGB and depth videos in multi-streams. The motivation behind our proposed work is to take the full advantage of RGB and depth videos simultaneously using multi-stream fusion at decision level. Another reason of motivation of our proposed work is to achieve the benefit of both automatic and hand-crafted features learning using multi-streams fusion at the decision level. Hence, our proposed approach earns the feature automatically using ST3DCNN in first and second streams and manual features from two motion templates such as MHI and MEI from RGB-D data parallelly for classification using 2DCNN in the last four streams. Therefore, this chapter proposes the fusion of six-stream for activity recognition using RGBD data. The workflow diagram of the suggested approach is shown in the figure 1.

Motivation

We are motivated to create a discriminative preserving activity recognition system after conducting a comprehensive survey on RGBD data. In this work, the main emphasis has been given on decision-level multi-stream fusion of space and time features to represent an activity; however, the color and depth-based (RGBD) activity recognition research employed on spatial streams of RGB and temporal stream of depth data (Ma et al., 2018). Some other researchers (Srihari et al., 2020) used two spatial streams and two motion streams on RGBD. Indeed, two-stream space-time fusion could be insufficient to capture discriminative spatial and temporal information from RGBD data. Thus, the fusion of more than two space and time streams is necessary to address inter and intra-class variations of activity problems. To implement the addressed issues, we have proposed a novel six-stream CNN fusion based on RGBD data with two spatial streams and four temporal streams. The first two spatial streams are fed by RGB and depth data directly, while the rest four streams use temporal features generated with MHI and MEI from RGB videos and depth videos.

To skip the segmentation part is another objective to carry out this work, which led to the motivation of the development of learning robust and accurate features directly from the raw data and deep learning has allowed learning these space and time features automatically. Irrespective of the hand-crafted features, it is essential to learn those features using deep neural networks (Khaire et al., 2018; Verma, Sah, & Srivastava, 2020).

The presented approach learns discriminant information from raw data and addresses the intra-class dissimilarity and inter-class similarity for RGBD efficiently. Besides feature learning, another emphasis is on decision level fusion to eliminate the drawbacks of the traditional features level fusion method.

Author Contribution

The overall contribution of our proposed work is given in below steps:

- This chapter introduces a vision based novel RGB and depth based human activity recognition framework.
- Although, there are many HAR systems are available now-a-days and most of them use two spatial streams and two temporal RGBD streams which limits the performance of the any HAR system. To address this problem, we have introduced four RGBD temporal streams along with two spatial streams.
- After validating on popular human activity recognition dataset such as UTD-MHAD, our proposed model as (depicted in figure 1) may be deployed in real time.

Chapter Organization

The rest portion of the work is organized in five sections where each section covers the different aspect of the work: The relevant review of literature is discussed in the section 2. In section 3 the detail explanation of the suggested approach has been discussed. The experimental work, results and discussion have been shown in the section 4. The work done in this chapter has been concluded with some future directions in the last section 5.

Challenges and Research Gaps

- Data collection through various sensors like Kinect depth, accelerator, and gyroscope sensors is the indivisible task for automatic human activity recognition, and then the systematic arrangement of the sensors is a critical issue that may affect the results. In addition, when faulty sensors are used, we don't know if the captured data is correct.
- Recognizing multi-person activity is difficult if the data capturing device (sensor) embed in a public or home environment where multiple people are present. Similarly, group activity recognition at the same time performed by a person is also a challenging task.
- Choosing an appropriate classification algorithm is also a difficult task because different algorithms have varying degrees of computational and time complexity, as well as varying degrees of accuracy.
- Since most activity recognition methods have been tested on standard datasets, their performance may vary when used in real-time applications.
- Outdoor activity recognition necessitates the use of a global positioning system (GPS) to track the location; however, indoor activity recognition is impossible without the deployment of a sensor, which, once again, creates a problem due to the presence of multiple people.

- Relevant feature selection and extraction from heterogeneous data is a challenging task.
- Last but not least, the classification models like deep neural network and traditional machine learning classifier frequently suffers overfitting and under-fitting problem which is again a challenging problem during developing recognition systems.

RELATED WORK

A large number of activity recognition techniques have discussed by the researchers in previous literature using RGB-D data. In this section, we have considered only those activity recognition approaches that utilize 1) CNN specially 3DCNN for activity recognition and 2) motion templates such as MHI and MEI. In addition to that, we have also reviewed multi-stream approaches for activity recognition.

Activity recognition using CNN (Krizhevsky et al., 2012) became very popular after the existence of convolutional neural network. Multi-stream CNN has also been used frequently in the literature for activity recognition task. For example, Karpathy et al. (Karpathy et al., 2014) proposed a two-stream multi-resolution CNN, in which first is fovea and second is context stream respectively. The consecutive frames are fed as an input to these streams. Then, several fusion methods like early, late and slow fusion are used to fuse the information extracted from both streams. In order to validate the approach a large-scale sports 1M dataset is used, which consists of more than one million sports videos from YouTube and have been categorized into 487 different activity classes. Recently, Verma *et al.* (Verma, Singh, Mandoria et al, 2020) proposed two-stage human activity recognition using 2DConvNet. The 2DConvNet learn spatial features efficiently from the 2D-images but it is not efficient to hold the temporal dependencies between the frames.

However, most of the approaches used in the literature consist of two separate streams one is for temporal and another for spatial information. To deal with this issue Hou *et al.* (Hou et al., 2020) proposed tube-CNN (T-CNN) using the descriptive power of end-to-end 3DCNN. In this approach, the fixed sized video input is fed into the Tube Proposal Network (TPN) and groups of proposal are achieved. Next, the tube proposals are connected to each other based on their action scores and finally, Tube-of-Interest (TOI) pooling is applied to combine the proposals to find the fixed sized features for action classification. The proposed approach in (Hou et al., 2020) has been validated on four datasets such as UCF-Sports, J-HMDB, THUMOS-13 and THUMOS-14' and achieved start-of-the art results.

As per the above discussion we deduced that the 3DCNN performed well for activity recognition task using 3D video data. But the existing 3DCNN has two major requirements 1) it requires fixed size of input video such as height and width 2) fixed number of input frames in input stream. Therefore, to tackle these issues Wang *et al.* (Wang et al., 2017) proposed two-stream 3D ConvNet fusion approach that deals with the input videos sequences using varying size and length. Firstly, spatial-temporal pyramid pooling strategy is proposed to deal with first issue of variable size input shots such as height and width, secondly Long Short-Term Memory (LSTM) and CNN-E model are used to deal with variable length input sequence to classify human actions. The proposed method given in (Wang et al., 2017) is validated on three challenging datasets such as UCF-101, HMDB-51 and ACT datasets and has given very promising results over state-of-the art results. Since, the present approaches are very computationally expensive due to the heterogeneous and complex structure of the network, therefore another video based activity recognition approach is proposed by Liu *at el.* (Liu et al., 2018) To address this issue they proposed a new real time Temporal Convolutional 3D Network (T-C3D) network that not only learn the temporal evaluation of visual features but also learns the absolute dynamics of the whole video sequence. The proposed approach given in (Liu et al., 2018) is trained and tested on the two benchmark datasets such as UCF-101 and HMDB-51, and generates more accuracy by a factor of 5.4% on the mentioned state-of-the art results. The execution speed of the proposed method (Liu et al., 2018) is twice as compared to the method given in the state-of-the arts. Hence, depending on the above discussion of the remarkable performance of 3DCNN for activity recognition in videos using 3D convolution features we have also used 3D-convolution features from both RGB and depth sequences for the later classification by SVM. The combination of 3DCNN+SVM has an extraordinary advantage over the previous methods where only 3DCNN has been used in the literature.

Activity recognition using motion templates like motion history and motion energy images have become very popular, because they both represent the motion information of a complete video sequence in a single motion template. To assess the discriminative power of these motion templates, Verma *et al.* (Verma, Sah, & Srivastava, 2020) computed both MHI and MEI templates in two different streams using the RGB video sequences for activity recognition. Then, they obtained the softmax scores using 2DCNN classification for the fusion at decision level. The proposed method is validated on three multi-model datasets such as NTU-RGB+D120, UTD-MHAD, CAD-60 and achieved 96.5% accuracy on UTD-MHAD dataset. The precision and recall value for CAD-60 dataset are 93.2% and 91.9%. The NTU-RGB+D120 dataset is evaluated for cross subject accuracy 76.7% and cross-setup accuracy 77.9%. However, MEI is enough capable to learn the temporal information rather than learning of both spatial-temporal information and can effectively train the deep-CNN. For example, Abdelbaky*et al.* (Abdelbaky & Aly, 2020) proposed multiple short-term motion energy images (ST-MEI) and principal component analysis network (PCANet). The proposed PCANet is an unsupervised learning method that learns the hierarchical local motion dynamic features from the proposed ST-MEI for later classification by linear SVM. The proposed approach (Abdelbaky & Aly, 2020) is validated using LOVO cross validation on the three datasets such as UCF sports, KTH and Weizmann. The obtained results show the effectiveness of the proposed method over mentioned state-of-the art methods.

Table 1. Some of art-of-the-state Approaches with their datasets and accuracy

Name	Approach	Datasets	Accuracy	Classes	Applications
Xia et al. (Xia et al., 2012)	Histogram of 3D Joints (HOJ3D) +LDA+HMM	UTKinectAction3D	90.92%	10	Indoor Activity Recognition
Li et al. (Li et al., 2010)	Action Graph + Bag of 3D points + Sampling	MSR Action3D	94.2%	20	Video Activity Recognition
Chen et al. (Chen et al., 2015)	Depth Motion Map	UTD-MHAD	79.1%	27	Activity Recognition
Shahroudy et al. (Shahroudy et al., 2016)	Part-Aware LSTM Network	NTU RGB+D	62.93% (cross-subject), 70.27 (cross-view)	60	Human Action Recognition
Ji et al. (Ji et al., 2012)	3DCNN	TRECVID2008, KTH	69.93%, 90.2%	3, 6	Action Recognition
Zhao et al. (Zhao et al., 2019)	3DSTCNN + 3D Skeleton Image + SVM	UTKinectAction3D and MSRActio3D	94.5%, 94.15%	10, 20	Activity Recognition
Niu et al. (Niu & Suen, 2012)	CNN-SVM	MNIST dataset	99.81%	10	Handwritten digits classification
Xue et al. (Xue et al., 2016)	CNN-SVM + Data Augmentation	NHI-ME	92.72%	2	Medical Disease Classification
Sargano et al. (Sargano et al., 2017)	CNN + Hybrid Model (SVM+KNN)	UCF Sports and KTH	91.47%, 98.15%	10, 6	Human Activity Recognition
Ma et al. (Ma et al., 2018)	CNN	JHMDB, MPII Cooking	76.9%, 70.3%	21, -	Activity Recognition

Continued on following page

Table 1. Continued

Name	Approach	Datasets	Accuracy	Classes	Applications
Khaire et al. (Khaire et al., 2018)	MHI +Depth Motion Map (DMM) + Skeleton Image + VGG-16	CAD-60, SBU Kinect Interaction and UTD-MHAD	93.06, 96.67, and 95.11	60, 8, and 27	Human Activity Recognition
Bulbul et al. (Bulbul et al., 2019)	3D-Motion Trail Model + MHI + SHI+ Gradient Local Auto co-relation (GLAC)	MSRAction3D, UTD-MHAD and DHA	99.34%, 89.5% and 99.1%	20, 27 and 23	Action Recognition
Sharma et al. (Sarma et al., 2020)	3DCNN+ 2DCNN based optical flow guided Motion Template	Palm's Graffiti Digits	99.2%	10	Digit Recognition
Zhou et al. (Cong & Zhang, 2020)	Deep Multi-Model Feature Fusion Approach	UCF Youtube, HMDB-51, Hollywood 2 and Olympic sport	81.93%, 85.37%, 74.28% and 76.25%	101, 51, 12 and 16	Activity Recognition
Karpathy et al. (Karpathy et al., 2014)	Multiresolution foveated ST-CNN	Sports 1 M dataset	63.9%	487	Sports Action Recognition
Verma et al. (Verma, Singh, Mandoria et al, 2020)	Corse and Fine level Classification + HOG + Random Forest + 2DCNN	UTKinectAction3D dataset	97.88%	10	Human Action Recognition
Hou et al. (Hou et al., 2020)	End-to-End 3DCNN + T-CNN	UCF Sports, J-HMDB and UCF-101	95.7%, 78.67% and 77.9%	10, 21, 101	Action Recognition
Wang et al. (Wang et al., 2017)	Two-Stream 3DCNn + Spatial Temporal Pyramid Pooling CNN (STPP-CNN) + LSTM	UCF101, HMDB-51	94.2% and 70.5%	101, 70.5	Action Recognition
Wang al.(Imran & Kumar, 2016)	MHI +Depth Motion Map (DMM) + CNN	UTD-MHAD	91.2	27	Human Activity Recognition

In addition to that, many previous approaches used multi-streams network to learn space and time features for activity recognition, for example the method in (Khaire et al., 2018) proposed a three-stream CNN network for activity recognition using three different modalities like RGB, depth map and skeleton joints information. The proposed approach generates MHI from the RGB videos in the first stream. The first stream is trained on these MHI by the VGG-F pre trained network and the corresponding action scores have been obtained for fusion with others streams at decision level. The proposed approach in (Khaire et al., 2018) is implemented on three challenging RGB-D datasets namely, CAD-60, SBU Kinect interaction and UTD-MHAD datasets. The proposed approach achieved 93.06% precision and 90.0% recall on the CAD-60 dataset, 96.26% accuracy on SBU Kinect interaction dataset and 94.60% accuracy on the UTD-MHAD dataset. After motivated from performance of motion templates in above discussions we have also used these motion templates MHI and MEI to obtain motion information from both RGB video and depth sequences. These MHI and MEI have been used in third, fourth, fifth, and sixth streams of our proposed methodology to obtain the scores of each test activity. Moreover, deep multi-stream CNN approaches (Chen et al., 2015; Cong & Zhang, 2020; Sarma et al., 2020; Verma, Sah, & Srivastava, 2020; Zhao et al., 2019) have substantially outperformed over single streams methods (Aggarwal & Ryoo, 2011; Ji et al., 2012; Xia et al., 2012). Therefore, this chapter also suggests multi-streams CNN fusion for spatial-temporal learning using multiple streams.

In (Verma & Singh, 2021) Verma et al. presented a multi-model approach using RGBD and skeleton data. They used 3DCNN to process RGBD data and an RNN network especially LSTM network to process skeleton data. The presented approach in (Verma & Singh, 2021) has been validated on UTKinectAction3D and MSRDailyActivity3D datasets and achieved 96.5% and 85.94% accuracies respectively. In (Verma et al., 2019) Verma et al. proposed a review on supervised and unsupervised based machine learning techniques for human activity recognition. Similarly, in (Banjarey et al., 2021) Banjarey et al. also performed survey on HAR problem using sensors and deep learning methods. In (Challa et al., 2021) Challa et al. used CNN-BiLSTM network for human activity recognition problem and validated the proposed approach on benchmark WISDM, UCI-HAR, and PAMAP2 datasets and achieved 96.05%, 96.37%, and 94.29% accuracies respectively. In (Dewangan et al., 2021) Dewangan et al. proposed vision-based lane region detection (VLDNet) network for intelligent vehicle detection system. In (Dewangan & Sahu, 2021) Dewangan et al. suggested a semantic segmentation-based CNN for road detection system. In (Dua et al., 2021) Dua et al. proposed a novel multi-input based human activity recognition approach and validated on three publicly available datasets such as UCI-HAR, WISDM and PAMAP2 and achieved 96.20%, 97.21% and 95.27% recognition accuracies.

PROPOSED APPROACH

In this proposed work to recognize human activities, a novel six-stream CNN fusion approach using RGB-D data has been presented. The objective of the stated approach is to take the full advantage of RGB and depth data simultaneously through the multiple streams. The first two streams are used to learn both space and time 3D convolutional features form both RGB and depth data using a spatial-temporal 3D-convolutional neural network (ST-3DCNN). Then, using the extracted 3D convolutional features, a support vector machine (SVM) is trained to estimate the score of each activity.

The remaining four streams are only used to learn temporal features, with the third and fourth streams extracting a powerful MHI motion template from RGB and depth data, respectively. Similarly, in the fifth and sixth streams, a MEI motion template is extracted from the RGB and depth data. The temporal features from the MHI and MEI extracted from the last four streams are then learned using a 2DCNN, followed by a SVM to estimate the class scores of each test activity. Finally, to obtain the system's overall accuracy, a decision level fusion scheme, specifically a weighted product model (WPM), is used to fuse the scores obtained from all streams for all activity classes.

Overall, the proposed work is divided into three levels. In the first level, space-time features are learned from RGBD data from all six streams; in the second level, Next, the features taken from all six streams are used to train an SVM; and in the final level, the score matrix obtained from all six streams is fused for decision making. The spatial-temporal features learning from all the streams using the RGBD data have been discussed in the subsequent sections.

Space Time Features Learning From RGBD Using ST3DCNN

The ST3DCNN is a spatial-temporal 3D-Convolutional neural network discussed in (Ji et al., 2012) that can acquire the information along time and space dimension at the same time. The convolutional 3D features may be obtained by convolving the 3D filter over the 3D volume input data. The 3D volume data is a sequence of frames stacked over and over again. The temporal information is learned along with depth dimension and the different feature map are obtained from multiple adjacent frames from the previous layer. To extract both lower and higher dimensional features, multiple convolutional layers are used. Thus, to increase the number of feature maps in the network large number of convolutional layers are used. Therefore, a convolutional 3D is achieved by convolving the 3D filter on the 3D input cube. Finally, the value in the ith layer's j^{th} feature map at a position (x, y, z) is given in (1)

$$v_{i,j}^{x,y,z} = \tanh \left(b_{ij} + \sum_m \sum_{a=0}^{A_{i-1}} \sum_{b=0}^{B_{i-1}} \sum_{c=0}^{C_{i-1}} w_{ijm}^{abc} v_{(i-1)m}^{(x+a)(y+b)(z+c)} \right) \tag{1}$$

Where w_{ijm}^{abc} is the (a, b, c)th value of the kernel of mth feature map in the previous layer.

Figure 2. Architecture of ST3DCNN for both RGB and depth video sequences

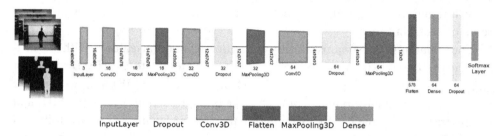

The ST3DCNN contains three convolutional and three pooling layers. The ST3DCNN model takes the predefined size of input frame, thus every frame is changed to a new size of [80, 80] with the help of Open CV library written in python. Because every video in the dataset has different number of frames, standardizing all videos of the dataset to equal number of frames is challenging task, hence we stacked fixed number of frames along with depth dimension. To carry out the experiment, 16 frames have been taken for the depth dimension. Thus, the input cube which is fed to the ST3DCNN model has the dimension [16×80×80]. Initially, the first convolution layer (C1) process the input cube and then first max pooling (ML1) layer process the output of C1. Meanwhile, 30% dropout is used between the ML1 and C1. The model uses another set of convolutional layers (C2) and maxpooling layer (ML2) with 40% dropout between C2 and ML2. Similarly, the model uses third set of convolutional layers (C3) and maxpooling layer (ML2) with 60% dropout between C3 and ML3. After the third max pooling layer, the model uses a fully connected layer (FC) to find the feature vector. The model has one dense layer with 64 neurons after the fully connected layer. At the end of our suggested model, a softmax layer is used for the output. Just before the softmax layer, the final dropout layer with a 50% dropout amount is used. The model uses 16, 32 and 64 filters of size (3×3×3) each in the C1, C2 and C3 layers respectively. The network employs a different down sampling size in each maxpooling layer such as (1×2×2), (2×3×3) and (3×3×3) in MP1, MP2 and MP3 layers respectively. The same network parameters have been used to process both RGB and depth video sequences of UTD-MHAD, MSRDailyActiviy3D and UCF Sports Activity datasets. Figure 2 depicts the ST3DCNN architecture for learning space-time features from RGB and depth video sequences.

Motion History Images (MHIs) From RGBD Data

Motion history image (MHI) is one of the easiest and simplest ways to represent the motion present in the video sequences. MHI specify the entire motion of a video sequence in a single frame. The MHI represents the scalar value image of a video sequence in which the pixel that has recent motion is the brightest pixel whereas the pixel with earlier movement represents the past motion. MHI uses the temporal density of pixels to convey the flow of motion. The zero-intensity value of a MHI pixel indicates that no movement has been recorded in the pixel during the frame history. The MHI can be represented by H_τ (x, y, t) and may be calculated by the update function Ψ (x, y, t) given in (2)

$$H_\tau\left(x,y,t\right) = \{{}^{\tau}_{\max\left(0,H_\tau\left(x,y,t\right)-\delta\right),otherwise} \tag{2}$$

Where the update function Ψ (x, y, t), represents the presence of the motion in the recent frame of the video sequence. Variable (x, y) represents the position of the pixel in the image; δ denotes the decay parameter and t represent the time. The time extent of the MHI has been described by τ. An update function Ψ (x, y, t) is used for every frame of the video. The value of the decay parameter δ has been obtained using experimentation and ranges from 15 to 20 for the different datasets in our experimentation. The update function Ψ (x, y, t), takes the value by the frame subtraction method using threshold given in (3)

$$\Psi\left(x,y,t\right) = \{{}^{1,if\,d\left(x,y,t\right)\geq\epsilon}_{0,otherwise} \tag{3}$$

Where $d\left(x,y,t\right)$ is the frame differencing defined in (4)

$$d\left(x,y,t\right) = \left|I\left(x,y,t\right) - I(x,y,t \pm \Delta\right| \tag{4}$$

Where $I\left(x,y,t\right)$ represents the intensity pixel value of the frame at time t. In our work, the value of Δ is 1, indicating that we used consecutive frame differencing to generate the motion history images (MHI) for dataset. In this work, MHI has been obtained from both RGB and depth video sequence in third and fourth streams respectively. Since, the depth sequences only provide the distance information, MHI can't be obtained directly from the depth video sequence thus, in order to find the MHI corresponding to the depth video sequence in the fourth stream; we first converted the depth video into the sequence of color images using different colormaps. The detailed generation of motion history image is given in Algorithm 1.

Algorithm 1: MHI Generation from Videos

```
Input: A directory of .avi or depth videos
Start; inc =10;
Function VideoReader (readpath);
        return videoobject;
nFrames ← videoobject.NumberofFrames;
Function ReadImage(background.jpg);
        return back_frame;
[x, y] ← size(back_frame);
α =zero(x, y);
Function SaveMotionHistoryImage(writepath);
        return MHI;
for k ← 1...........nFrames do
        Iₜ    = Read back_frame;
        Iₜ₋₁ = Read kᵗʰ frame;
        βₜ = absolute_difference (Iₜ, Iₜ₋₁);
    for i ← 1...........x do
        for j ← 1...........y do
                If    βₜ < 50    then
                        α ← α ₜ - 0.15;
                else
                        α (i, j) ← inc
                end
        end
    end
Increment inc by 1;
call SaveMotionHistoryImage();
end
Output: A directory of Motion History Images
```

Motion Energy Images (MEIs) From RGBD Data

Motion energy image (MEI) (Blank et al., 2005) is a binary cumulative motion image. MEI essentially encodes all motion information in a single binary image. The binary image describes region of motion in the image regardless the direction of the motion. MEI is accomplished through the use of either frame differencing or the background subtraction method. Background subtraction is used for data where the background is static, and the silhouette of the human body is extracted, providing sufficient information about the human body's movement. The frame differencing method, on the other hand, is useful when the input data has a variable background. In this study, we used the frame differencing method to determine the MHI. Suppose, I (u, v, t) & I (u, v, t+1) are two successive frames of a video sequence at a point of time t & t+1. The absolute difference D (u, v, t) between these two frames can be calculated using (5)

$$D \ (u, v, t) = \left| I\left(u, v, t+1\right) - I\left(u, v, t\right) \right| \tag{5}$$

The D (u, v, t) image is transformed to binary image E (u, v, t) by setting the threshold value to represent the existing motion region. The MHI is defined in (6).

$$E \ (u, v, t) = \bigcup_{i=0}^{\tau-1} D(u, v, t - i) \tag{6}$$

Where τ is the temporal range of the movement. We used the frame differencing method to obtain the MEI from the RGB and depth sequences in the fifth and sixth streams of our proposed work. The detailed generation of motion energy images is given in Algorithm 2.

Algorithm 2: Motion Energy Image Generation from Videos

```
Input: A directory of .avi or depth videos
Start;
Function VideoReader (readpath);
        return videoobject;
nFrames ← videoobject.NumberofFrames;
Function ReadImage(background.jpg);
        return back_frame;
[x, y] ← size(back_frame);
α =zero (x, y);
Function SaveMotionEnergyImage (writepath);
        return MEI;
for k ← 1..........nFrames do
        I_t    = Read back_frame;
        I_{t-1} = Read k^{th} frame;
        β_t = absolute_difference (I_t, I_{t-1});
    for i ← 1..........x do
        for j ← 1..........y do
                If    β_t < 60    then
                        α[i, j] ← α [i, j]  - 0;
                else
                        α[i, j] ← α [i, j]  - 255;
                end
        end
    end
call SaveMotionEnergyImage();
end
Output: A directory of Motion Energy Images
```

Temporal Features Learning Using 2DCNN

A 2DCNN is a type of deep neural network that is specifically designed for grid-type input data such as image classifications and other similar tasks. The CNN is inspired from the biological structure of the human brain where the connectivity pattern of the neurons is similar to the animal visual cortex (Krizhevsky et al., 2012). CNN is a sequence learning classifier that learns features through convolutional operations. The CNN model consists of a series of sequential layers stacked one on top of the other, with each layer's function being to transform the output of the previous layer into the input for the next layer in the sequence. In 2DCNN, there are primarily four types of layers that are used to build the CNN model. The layers may be Convolutional layer, pooling or subsampling layer, dropout layer and fully connected layer. Apart from these layers, another layer known as the softmax layer is used for classification. We arrange the layers one over another to build the CNN model. In comparison to other classification techniques, the amount of preprocessing required for training the CNN model is less. The most imperative aspect of CNN is the automatic learning of features from the input data. In this work, we used 2DCNN model to learn the features from the MHI and MEI motion templates corresponding to RGB and depth video sequences in third, fourth, fifth and sixth streams of our proposed approach. Our proposed 2DCNN model is trained with a loss function based on categorical cross entropy (CCE). Figure 3 depicts the architecture of our proposed 2DCNN model for feature learning from MHI and MEI corresponding to RGB and depth sequences.

CCE Based 2DCNN: Let us consider a network having C distinct output values, one of which corresponds to each activity sequence. The categorical cross entropy (CCE) loss function is then defined in (7)

$$\text{CCE} = -\frac{1}{N}\sum_{j=1}^{N}\sum_{k=1}^{C}l_{yi \in Ck}\text{logProb}_{model}[_{yi \in Ck}] \tag{7}$$

In order to learn the temporal features from both MHI and MEI streams extracted from the RGB and depth video sequences, a 2DCNN model has been used. The proposed 2DCNN model contains three convolutional layers and three max-pooling layers. To prepare the input for the 2DCNN model, simple preprocessing was carried out in which the frames of the input videos were rescaled to fixed height and width [80, 80] dimensions using the OpenCV library in Python. Thus, the size of the MHI and MEI motion templates used as input to the 2DCNN is [80(height) ×80(width)]. This input is first processed by C1 (first convolutional layer), followed by P1 (first max-pooling layer). The output of the first maxpooling layer (P1) fed into the next convolutional layer (C2) followed by second maxpooling layer (P2). Similarly, the network also contains the third convolutional layer (C3) followed by the third maxpooling layer (P3). Then, a flatten layer is used to obtain the one-dimensional features. Following the flatten layer, two fully connected dense layer have been used with 128 neurons and 64 neurons respectively. The network contains 8, 16 and 32 features map in the first, second and third convolutional layers (C1, C2 and C3) respectively. The size of the kernel and stride of each convolutional layer are set to (3x3) and 1 respectively. For the first two maxpooling layers (P1&P2), a (2x2) down sampling is used, and for the third maxpooling layer, a (3x3) down sampling is used

(P3). Before each maxpooling layer, a dropout layer has been used with an amount of 0.2, 0.3 and 0.6 respectively to prevent the over fitting. During the experiment, the same network parameters have been used for feature learning from the MHI and MEI streams for UTD-MHAD, MSRDailyActivity3D and UCF Sports Activity datasets. Figure 3 depicts the architecture of the proposed 2DCNN.

Figure 3. Architecture of proposed 2DCNN model used for training and testing the MHI and MEI motion images.

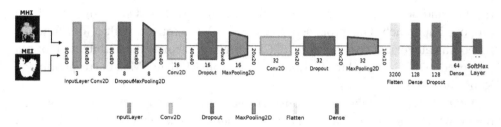

Activity Score Generation Using SVM

SVM is a classifier build on kernels (Chapelle et al., 2020). The classifier's main objective is to map the input samples into the higher dimensional feature space and insert a hyperplane that distinctly classifies the input samples. SVM uses different kernels to perform both linear and non-linear classification. SVM kernels include linear, radial bases function (rbf), polynomial, etc. In our work, SVM classifier is trained using the training features extracted from each parallel stream. In the first two streams, SVM classifier is used to obtain the activity score from RGB and depth sequences using spatial and temporal feature learning with the ST3DCNN model. Similarly, in the remaining four streams, SVM classifier is used to obtain the activity score from the same RGB and depth video sequences using two motion templates such as MHI and MEI. During the testing period, the testing features corresponding to ST3DCNN, MHI, and MEI have been extracted and are applied to the trained SVM model for all parallel streams. The predicted probability score for each test activity is then computed. Finally, the obtained probability scores for each test activity in each stream are fused together to obtain the final score of each activity class.

Decision Level Score Fusion (DLSE) Approach

Finally, the decision level fusion approach has been used to predict the probability score of each test activity in each stream streams. The DLSF approach takes classification results from individual streams and processes them to produce more robust results after combining them. In DLSF, there are two major approaches. In the first approach, an additional training is required (supervised) and, in the second approach there is no training required (unsupervised). The probability score or output of the classifiers is inserted as an input to the training model in the supervised approach. Unsupervised approaches, on the other hand, do not require any additional training. There are some unsupervised approaches to integrate the classifiers are: Average Method, Borda Count and Weighted Product Model (WPM). In this chapter, we used WPM as a DLSF approach.

Weighted Product Model (WPM)

The weighted product model, which is based on the multi-criteria decision making (MCDM) method, is a well-known decision level fusion strategy. The main challenge in MCDM is combining the probability scores acquired from the various streams. It is same as the weighted sum model (WSM) method (Dhanisetty et al., 2018). The major difference between WPM and WSM is that WPM uses multiplication rather than summation. Here in WPM, the predicted probability scores are combined together in order to find the most appropriate class of an activity.

In this paragraph the brief description of the WPM method is given. Suppose there are p number of decision statements and q number of alternatives for a specific MCDM problem and r_j indicates the relative weights for the jth statement and s_j indicates the performance of the alternative A_k, which is calculated in the statement j. Then, the WPM is defined in (8) as follows:

$$P(A_k) = \prod_{j=1}^{p} \left(a_{s_j} \right)^{r_j}, \text{ for k} = 1, 2, 3\ldots q. \tag{8}$$

Let S1, S2, MHI$_{RGB}$, MHI$_{Depth}$, MEI$_{RGB}$, MEI$_{Depth}$ denotes the scores obtained using independent streams of ST3DCNN, MHI and MEI for RGB and Depth data respectively. The scores obtained from the separate streams work as decision criteria while the alternatives define the number of activities. Every stream generates a score matrix vector having the same number of activities. Based on this structure, the best suitable alternative activity (having the maximum value of P (A_k)) gets selected. The weight values such as W_1, W_2, W_3, W_4, W_5 and W_6 were set to 1 for the dataset.

Here, if all the weight values become same then the weighted product model approach works as product rule. The score fusion using WPM method is defined as follows in (9).

$$WPM_S = \max [(S1^{w1}) \times (S2^{w2}) \times (MHI_{RGB}^{W3}) \times (MHI_{Depth}^{W4}) \times (MHI_{RGB}^{W3})$$
$$\times (MHI_{Depth}^{W3})] \tag{9}$$

The proposed approach has been trained using Algorithm 3 and given in the subsequent section.

Algorithm 3: Proposed Model training using RGBD data, MHI and MEI

Input: A pair of RGB and Depth Video streams and Ground truth $\left(Video_{RGB}, Video_{Depth}\right)_{i=1}^{N}$ and G_T

Start;

$[MHI_{RGB} MHI_{Depth}]$ = Estimate MHI $(Video_{RGB}, Video_{Depth})$;

$[MEI_{RGB} MEI_{Depth}]$ = Estimate MEI $(Video_{RGB}, Video_{Depth})$;

for *Epoch1* ← *1............* N_E **do**

    for *Epoch2* ← *1............N* **do**

        Extract Feature Maps $(Video_{RGB}, Video_{Depth})$ using (1)

        Extract Feature Maps $(MHI_{RGB} MHI_{Depth})$ using *2DCNN*

        Extract Feature Maps $(MEI_{RGB} MEI_{Depth})$ using *2DCNN*

        Update parameters using loss function in (7)

        If *reducing loss* **then**

            Update the Feature map set F_m;

            Update ground truth G_T;

        else

            go to next *Epoch1*;

        end

    end

end

Function Predicted Probabilities (F_m, G_T);

      return scores;

An activity score generation using (8) and (9);

Output: A matrix of activity score

Figure 4. Samples of 27 activities in UTD-MHAD

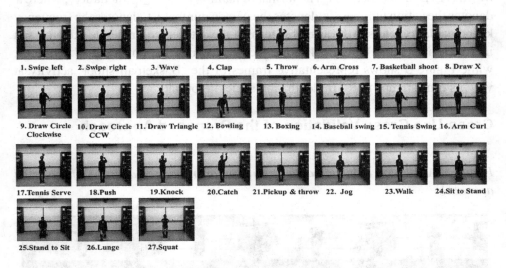

EXPERIMENTAL RESULTS

This work used UTD-MHAD (Chen et al., 2015) dataset to train and test our proposed approach. Kinect sensor has been used to capture the dataset. All the activities in the dataset are synchronized in all modality. In order to train the proposed network, we used Intel Core i7 8th generation processor on Ubuntu 16.04 long term support operating system with RAM capacity of 16 GB along with NVIDIA graphic support of 2GB. To implement CNN networks, we used Keras Deep learning framework with version 2.2.4, and for generating the MHI and MEI from the input RGB and depth videos, we used MATLAB version 18a.

UTD-MHAD Dataset

The UTD-MHAD dataset (Chen et al., 2015) was captured with the help of two sensors such as wearable and kinect sensor in 2015. There are 27 activities in the dataset carried out by 8 different persons (4 males and 4 females). Every person repeated each activity four times, leading to the 8 (subjects) × 27 (activities) × 4 (frequency) = 864 activities sequences. The dataset was captured in four modalities such as RGB videos, depth map, skeleton joint positions, and the inertial sensor data. Three channels were used to capture the data in which, RGB used one channel and one channel for inertial sensor data and one channel for both skeleton joints and depth maps. The action is synchronized in all modalities. The file format of RGB videos is (.avi), while depth, skeleton and inertial sensor data were recorded by MATLAB as a (.mat) file. The format of naming is "$a_i\_s_j\_t_k\_$modality", where a_i denotes the action number i, s_j denotes the person number j, t_k denotes the frequency k, and modality refers to the available form of the dataset (color, depth, skeleton and inertial sensor signal). Figure 4 shows pictorial representations of all 27 activities, while Figure 5 shows representative frames of the 'Basketball shoot' activity at various time stamps in RGB, depth, and skeleton format.

Cross-Use Cross Validation of the Proposed Approach

Figure 5. Allocation of human activity dataset for training and testing of proposed approach

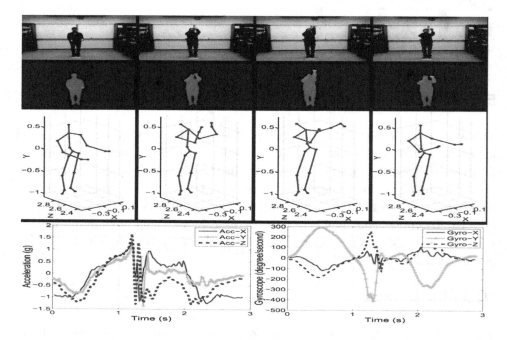

Figure 6. Sample frames of 'Basketball shoot' activity in UTD-MHAD dataset at different time stamp in three formats RGB, Depth and Skeleton.

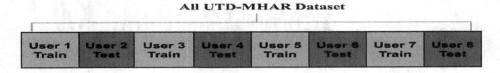

We adopted the same cross-user (CU) cross-validation method used in (Chen et al., 2015) for validation of our proposed methodology, in which users 1, 3, 5, 7 were used for training and users 2, 4, 6, 8 were used for testing. The allocation of the dataset into training and testing is shown in the figure 6. In cross user cross-validation, 4(user) $\times$ 27(activates) $\times$ 4 (frequency)= 432 activity sequences have been used in training corresponding to users 1, 3, 5, 7 and $4 \times 27 \times 4 = 432$ activity sequences have been used for testing corresponding to users 2, 4, 6, 8. During the testing phase, the class probability scores of each activity sequence have been recorded for all streams.

Figure 7. Training curve of our proposed approach

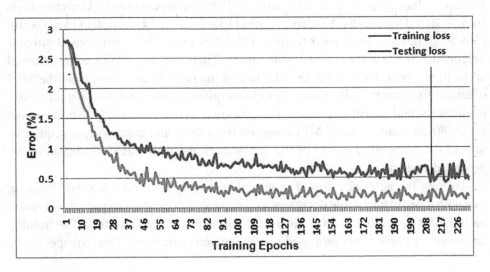

Figure 8. Accuracy curve of our proposed approach

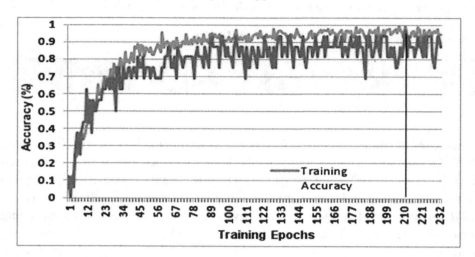

Using ST3DCNN and SVM independently in the first and second streams of our proposed system, we acquired probabilities scores from RGB and depth video sequences during the implementation phase. Different input cube sizes combinations from RGB videos and depth videos, such as (13×50×50), (14×50×50), (15×50×50), (16×50×50), have been tried to the ST3DCNN in the first and second streams independently for activity score generations. Then, the best scores were obtained at the input size (16×50×50). In order to find the probability scores from the MHI obtained from both RGB and depth video sequences, we used (80×80) input size in the third and fourth streams of our proposed system. Similarly, we have kept same (80×80) input size of MEI obtained from RGB and depth videos sequences to find the probability scores of the activities in the fifth and sixth stream of the proposed system.

To train the networks for our proposed approach, we used a 6×10^{-4} learning rate with a decay factor that decreases as the number of epoch's increases. An Adam optimizer is used along with a categorical cross-entropy loss function. The training and accuracy curve of our proposed system is given in the figure 7 and 8 respectively.

Figure 9. Individual activity class accuracy for UTD-MHAD dataset using weighted product model method

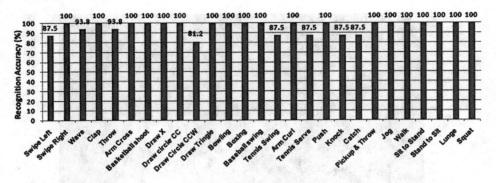

RESULTS AND DISCUSSIONS

This section describes the obtained experimental results using our proposed approach. Our proposed approach obtained 96.6% recognition accuracy of all 27 human activities of UTD-MHAD dataset. The confusion matrix of our proposed approach is given in the Fig 10. It is clear from the Fig 9 that 19 out of 27 activities of the dataset have been recognized with 100% accuracy. However, the activity 'Draw circle CCW' has less recognition accuracy which is 81.2% due to the similarity with 'Draw circle CC', 'Draw triangle' and 'clap' activities. Similarly, the activities 'knock' and 'catch' have less recognition accuracy compare to other activities due to higher confusion with each other. In addition to that the activities like 'Tennis swing' and 'Tennis serve' have also more confusion due to the activity performed by the external object (tennis). On the other hand, the recognition accuracy of 'Swipe left' activity is 87.5% and our model is recognizing 12.5% times 'swipe left' activity as 'wave' activity

Figure 10. Confusion matrix of our proposed approach on 27 activities of UTD-MHAD dataset

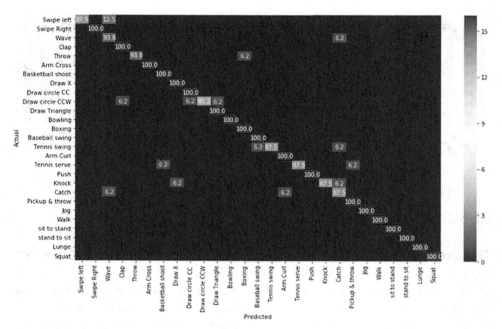

After the above results it is clear that our proposed approach is giving excellent performance and all the activities have good recognition accuracies except those have higher degree of confusion, which prove that our propose approach learns visual, temporal features and the relevant information from the input data in excellent way. Table 2 consists of state-of-art techniques along with achieved results on the UTD-MHAD data. It also compares the accuracy of our stated approach with the existing techniques. Table 1 is also showing that our proposed approach is giving the highest accuracy among the mentioned techniques.

Table 2. Performance of our proposed approach on UTD-MHAD dataset, compare to the state-of-art-approaches

Methods	Accuracy (%)
C.Chen*et al.*(Chen et al., 2015)	79.1
Bulbul *et al.* (Bulbul et al., 2019)	88.4
Imran J. *et al.* (Imran & Kumar, 2016)	91.2
Escobedo E. *et al.* [30	84.4
Khaire P. *et al.* (Khaire et al., 2018)	95.11
Our Proposed Approach	**96.60**

CONCLUSION

In this chapter, a fusion of six-stream CNN has been presented to solve the problem of recognition of human activities using RGBD data. The proposed approach is implemented in three different phases. In the first phase, both spatial and temporal features learning using ST3DCNN from RGB and depth video sequences along with temporal features learning using 2DCNN from MHI and MEI motion templates have been performed. Secondly, parallel SVM model has been applied to obtain the class scores of each test activity from the all streams. At last, the class probability scores obtained from each test activity have been fused at decision level using weighted product model (WPM) method. The method suggested in this work is enough capable to learn automatically high level space and time features directly from the input data, as well as from two motion templates such as MHI and MEI. Our proposed approach uses both RGB and depth data simultaneously and gives better performance rather than using individual modality separately. The proposed six-stream CNN approach has been validated on the UTD-MHAD dataset and is giving 96.60% recognition accuracy on 27 activities of the dataset. The obtained results show that there is an enhancement in the recognition rate by our proposed six-stream CNN approach over existing methods due to the addition of depth information along with color information.

REFERENCES

Abdelbaky, A., & Aly, S. (2020). Human action recognition using short-time motion energy template images and PCANet features. *Neural Computing & Applications*, *32*(16), 1–14. doi:10.100700521-020-04712-1

Aggarwal, J. K., & Ryoo, M. S. (2011). Human activity analysis: A review. *ACM Computing Surveys*, *43*(3), 1–43. doi:10.1145/1922649.1922653

Banjarey, K., Sahu, S. P., & Dewangan, D. K. (2021, April). A Survey on Human Activity Recognition using Sensors and Deep Learning Methods. In *2021 5th International Conference on Computing Methodologies and Communication (ICCMC)* (pp. 1610-1617). IEEE. 10.1109/ICCMC51019.2021.9418255

Blank, M., Gorelick, L., Shechtman, E., Irani, M., & Basri, R. (2005). Actions as space-time shapes. *IEEE International Conference on Computer Vision*.

Bulbul, M. F., Islam, S., & Ali, H. (2019). Human action recognition using MHI and SHI based GLAC features and collaborative representation classifier. *Journal of Intelligent & Fuzzy Systems*, *36*(4), 3385–3401. doi:10.3233/JIFS-181136

Challa, S. K., Kumar, A., & Semwal, V. B. (2021). A multibranch CNN-BiLSTM model for human activity recognition using wearable sensor data. *The Visual Computer*, 1–15. doi:10.100700371-021-02283-3

Chapelle, O., Vapnik, V., Bousquet, O., & Mukherjee, S. (2020). Choosing multiple parameters for support vector machines. *Machine Learning*, *46*(1-3), 131–159.

Chen, C., Jafari, R., & Kehtarnavaz, N. (2015). UTD-MHAD: A multimodal dataset for human action recognition utilizing a depth camera and a wearable inertial sensor. *IEEE International Conference on Image Processing*.

Cong, J., & Zhang, B. (2020). Multi-model feature fusion for human action recognition towards sport sceneries. *Signal Processing Image Communication*, *84*, 115803. doi:10.1016/j.image.2020.115803

Dewangan, D. K., & Sahu, S. P. (2021). Road Detection Using Semantic Segmentation-Based Convolutional Neural Network for Intelligent Vehicle System. In *Data Engineering and Communication Technology* (pp. 629–637). Springer. doi:10.1007/978-981-16-0081-4_63

Dewangan, D. K., Sahu, S. P., Sairam, B., & Agrawal, A. (2021). VLDNet: Vision-based lane region detection network for intelligent vehicle system using semantic segmentation. *Computing*, *103*(12), 1–26. doi:10.100700607-021-00974-2

Dhanisetty, V. V., Verhagen, W. J. C., & Curran, R. (2018). Multi-criteria weighted decision making for operational maintenance processes. *Journal of Air Transport Management*, *68*, 152–164. doi:10.1016/j.jairtraman.2017.09.005

Dua, N., Singh, S. N., & Semwal, V. B. (2021). Multi-input CNN-GRU based human activity recognition using wearable sensors. *Computing*, *103*(7), 1–18. doi:10.100700607-021-00928-8

Escobedo, E., & Camara, G. (2016). A new approach for dynamic gesture recognition using skeleton trajectory representation and histograms of cumulative magnitudes. *SIBGRAPI Conference on Graphics, Patterns and Images*. 10.1109/SIBGRAPI.2016.037

Hou, R., Chen, C., & Shah, M. (2020). *An end-to-end 3d convolutional neural network for action detection and segmentation in videos*. arXiv preprint arXiv:1712.01111.

Imran, J., & Kumar, P. (2016). Human action recognition using RGB-D sensor and deep convolutional neural networks. *International Conference on Advances in Computing, Communications and Informatics*. 10.1109/ICACCI.2016.7732038

Ji, S., Xu, W., Yang, M., & Yu, K. (2012). 3D convolutional neural networks for human action recognition. *IEEE Transactions on Pattern Analysis and Machine Intelligence*, *35*(1), 221–231. doi:10.1109/TPAMI.2012.59 PMID:22392705

Karpathy, A., Toderici, G., Shetty, S., Leung, T., Sukthankar, R., & Fei-Fei, L. (2014). Large-scale video classification with convolutional neural networks. *IEEE Conference on Computer Vision and Pattern Recognition*.

Khaire, P., Kumar, P., & Imran, J. (2018). Combining CNN streams of RGB-D and skeletal data for human activity recognition. *Pattern Recognition Letters*, *115*, 107–116. doi:10.1016/j.patrec.2018.04.035

Khoshelham, K., & Elberink, S. O. (2012). Accuracy and resolution of kinect depth data for indoor mapping applications. *Sensors (Basel)*, *12*(2), 1437–1454. doi:10.3390120201437 PMID:22438718

Krizhevsky, A., Sutskever, I., & Hinton, G. E. (2012). Imagenet classification with deep convolutional neural networks. *Advances in Neural Information Processing Systems*, *25*, 1097–1105.

Li, W., Zhang, Z., & Liu, Z. (2010). Action recognition based on a bag of 3d points. *IEEE Computer Society Conference on Computer Vision and Pattern Recognition-Workshops*. 10.1109/CVPRW.2010.5543273

Liu, K., Liu, W., Gan, C., Tan, M., & Ma, H. (2018). T-c3d: Temporal convolutional 3d network for real-time action recognition. *AAAI Conference on Artificial Intelligence*, *32*(1).

Ma, M., Marturi, N., Li, Y., Leonardis, A., & Stolkin, R. (2018). Region-sequence based six-stream CNN features for general and fine-grained human action recognition in videos. *Pattern Recognition, 76,* 506–521. doi:10.1016/j.patcog.2017.11.026

Niu, X. X., & Suen, C. Y. (2012). A novel hybrid CNN–SVM classifier for recognizing handwritten digits. *Pattern Recognition, 45*(4), 1318–1325. doi:10.1016/j.patcog.2011.09.021

Sargano, A. B., Wang, X., Angelov, P., & Habib, Z. (2017). Human action recognition using transfer learning with deep representations. *International Joint Conference on Neural Networks.*

Sarma, D., Kavyasree, V., & Bhuyan, M. K. (2020). *Two-stream Fusion Model for Dynamic Hand Gesture Recognition using 3D-CNN and 2D-CNN Optical Flow guided Motion Template.* arXiv preprint arXiv:2007.08847.

Shahroudy, A., Liu, J., Ng, T. T., & Wang, G. (2016). Ntu rgb+ d: A large scale dataset for 3d human activity analysis. *Proceedings of the IEEE conference on computer vision and pattern recognition.* 10.1109/CVPR.2016.115

Srihari, D., Kishore, P. V. V., Kumar, E. K., Kumar, D. A., Kumar, M. T. K., Prasad, M. V. D., & Prasad, C. R. (2020). A four-stream ConvNet based on spatial and depth flow for human action classification using RGB-D data. *Multimedia Tools and Applications, 79*(17), 11723–11746. doi:10.100711042-019-08588-9

Sung, J., Ponce, C., Selman, B., & Saxena, A. (2012). Unstructured human activity detection from RGBD images. *IEEE International Conference on Robotics and Automation.*

Verma, K. K., & Singh, B. M. (2021). Deep Multi-Model Fusion for Human Activity Recognition Using Evolutionary Algorithms. *International Journal of Interactive Multimedia & Artificial Intelligence, 7*(2), 44. doi:10.9781/ijimai.2021.08.008

Verma, K. K., Singh, B. M., & Dixit, A. (2019). A review of supervised and unsupervised machine learning techniques for suspicious behavior recognition in intelligent surveillance system. *International Journal of Information Technology,* 1-14.

Verma, K. K., Singh, B. M., Mandoria, H. L., & Chauhan, P. (2020). Two-Stage Human Activity Recognition Using 2D-ConvNet. *International Journal of Interactive Multimedia & Artificial Intelligence, 6*(2), 125–135. doi:10.9781/ijimai.2020.04.002

Verma, P., Sah, A., & Srivastava, R. (2020). Deep learning-based multi-modal approach using RGB and skeleton sequences for human activity recognition. *Multimedia Systems, 26*(6), 1–15. doi:10.100700530-020-00677-2

Wang, X., Gao, L., Wang, P., Sun, X., & Liu, X. (2017). Two-stream 3-D convNet fusion for action recognition in videos with arbitrary size and length. *IEEE Transactions on Multimedia, 20*(3), 634–644. doi:10.1109/TMM.2017.2749159

Xia, L., Chen, C. C., & Aggarwal, J. K. (2012). View invariant human action recognition using histograms of 3d joints. *IEEE Computer Society Conference on Computer Vision and Pattern Recognition Workshops*. 10.1109/CVPRW.2012.6239233

Xue, D. X., Zhang, R., Feng, H., & Wang, Y. L. (2016). CNN-SVM for microvascular morphological type recognition with data augmentation. *Journal of Medical and Biological Engineering, 36*(6), 755–764. doi:10.100740846-016-0182-4 PMID:28111532

Zhao, C., Chen, M., Zhao, J., Wang, Q., & Shen, Y. (2019). 3d behavior recognition based on multi-modal deep space-time learning. *Applied Sciences (Basel, Switzerland), 9*(4), 1–14. doi:10.3390/app9040716

Chapter 8

Multi-Input CNN-LSTM for End-to-End Indian Sign Language Recognition:
A Use Case With Wearable Sensors

Rinki Gupta

https://orcid.org/0000-0002-0060-5523
Amity University, Noida, India

ABSTRACT

Sign language predominantly involves the use of various hand postures and hand motions to enable visual communication. However, signing is mostly unfamiliar and not understood by normal hearing people due to which signers often rely on sign language interpreters. In this chapter, a novel multi-input deep learning model is proposed for end-to-end recognition of 50 common signs from Indian Sign Language (ISL). The ISL dataset is developed using multiple wearable sensors on the dominant hand that can record surface electromyogram, tri-axial accelerometer, and tri-axial gyroscope data. Multi-channel data from these three modalities is processed in a multi-input deep neural network with stacked convolutional neural network (CNN) and long short-term memory (LSTM) layers. The performance of the proposed multi-input CNN-LSTM model is compared with the traditional single-input approach in terms of quantitative performance measures. The multi-input approach yields around 5% improvement in classification accuracy over the traditional single-input approach.

DOI: 10.4018/978-1-7998-9434-6.ch008

INTRODUCTION

Sign language is primarily used by deaf and mute people for communication. A deaf person often finds it difficult to communicate with other people in form of written communication, since that would require knowledge of verbal languages such as English and Hindi, which are difficult to learn without the sense of hearing. On the other hand, sign language is a combination of hand gestures, facial expressions and body language in a well-defined semantics and lexicon that is easy to learn and utilize for a deaf person (Kudrinko, 2018; Gupta, & Kumar, 2021). However, sign language is not common among the hearing community making it difficult for a deaf person to communicate with them in signing. An electronic translator for converting sign language to spoken language could greatly enhance the communication between a sign language user and a non-signer. With this motivation, several researchers have reported about development of sign language recognition. Hundred and seventeen papers on recognition systems reported for 25 different sign languages over the past decade have been reviewed in (Wadhawan, 2021). Sign language recognition has been performed using non-wearable devices by capturing images, videos, depth or color related information, as well as wearable sensors such as flex sensors, motion sensors, electromyograms and even wi-fi signals. The reported research either explores the use of only the dominant hand during signing or both the hands. Moreover, the considered signs are either just static postures or information about the dynamic motion of hands is also captured. The review concludes that over 54% of work is dedicated towards reliably recognizing the posture of static hand(s) during an isolated sign (Wadhawan, 2021). Around 75% of the research was found to be focused on processing information from only the dominant hand. Use of vision-based approaches were found to be most popular, with 44% using camera and 23% using Kinect or leap motion sensor for data acquisition. The recognition may be carried out using machine learning and deep learning techniques.

A survey of literature on vision-based sign language recognition using deep learning techniques is presented in (Rastgoo, 2020). The authors report that isolated hand gestures without motion, such as those used for signing numerals and most alphabets may be classified using images. However, for identifying dynamic signs at word or sentence level, videos are required for data acquisition. Convolutional neural networks (CNN) and recurrent neural networks (RNN) have been found to be most prominently used for classification of hand and body posture and for hand tracking in dynamic signing. For instance, Wadhawan et al. evaluated 50 CNN models for determining hand posture from RGB images of 100 different signs from the Indian sign language (ISL) to achieve classification accuracy as high as 99.9% (Wadhawan, 2020). Hand pose has been determined using depth sensors for American sign language (ASL) in (Kolivand, 2021). First, the hand and forearm are segmented from the depth image of the hand sign. Thereafter, denoising and extraction of geometrical features is carried out to achieve classification accuracy of up to 96.78% with artificial neural network (ANN). Videos have also been used for recognition of dynamic signs. In (Masood, 2018), the authors used a combination of CNN and RNN to extract spatial and temporal features from video inputs to classify 46 gestures from Argentinean Sign Language with 95.2% accuracy. Despite their successful use for sign language recognition, some of the major challenges encountered with vision-based approaches are partial occlusion, poor illumination and background clutter (Rastgoo, 2020).

Wearable sensors have also been reported to be extensively useful for this task (Banjarey, 2021). A review of 72 studies on use of wearable sensors for sign language recognition reported in past eight years is presented in (Kudrinko, 2021). The review explores the use of surface electromyogram (sEMG) sensors for capturing muscle activity, inertial measurement unit (IMU) consisting of accelerometers and gyroscopes for capturing information about acceleration and angular velocity, and hand gloves equipped with flex sensors, fiber optic sensors and IMUs for measuring bending of fingers. Wearable gloves and armbands have also shown to be useful for recognizing hand postures. While around 64% of the studies were reported with conventional machine learning algorithms, such as support vector machine (SVM), k nearest neighbour (kNN) and random forest (RF), just around 3% of the reviewed studies employed deep learning models such as CNN and RNN. In (Khomami, 2021), the authors placed four sEMG sensors on the forearm and one IMU on the wrist to capture information for 20 commonly used signs from the Persian sign language. After pre-processing, activity detection and feature extraction, several machine learning classifiers such as kNN and SVM were employed and highest accuracy of 96.13% was obtained with the kNN classifier. In (Gupta, & Kumar, 2021), data was acquired using six sEMG and six IMU sensors on both forearms. The authors employed multi-label classification approaches such as binary relevance, label powerset and classifier chain designed using SVM classifier to attain average classification error as low as 2.73% for 100 ISL signs. Data gloves containing flex sensors and IMU have been designed to classify signs, such as those mentioned in (Dong, 2021; Yuan, 2021). Flex sensors have been used to measure the bending of the fingers, while the IMU provides information about the orientation and motion of hand. The authors in (Dong, 2021) employed techniques such as fusion of features extracted from 1D and 2D CNNs, stacking of CNN and LSTM to classify signs from ASL and Chinese sign languages (CSL) with up to 74.19% accuracy. In (Yuan, 2021), two arm-rings containing IMUs were used in addition to the hand gloves to record signs from ASL and CSL. Highest classification accuracy of 99.93% for static signs from ASL using a stacked CNN-LSTM architecture. Major challenges identified with the use of wearable sensors for sign language recognition are related to scalability and adaptability of the developed system (Kudrinko, 2021). Nevertheless, the review also reveals that most studies reported a ready-to-use device for sign language recognition, that are user specific.

Besides sign language recognition, studies related to hand gesture recognition are also useful in applications such as robot control (Dai, 2020), control of automated vehicles (Müezzinoğlu, 2021), and control of assistive devices such as wheelchair (Kundu, 2018) and prosthetics (Zandigohar, 2021). In (Dai, 2020), the authors designed an IMU-based device for controlling a mobile robot with dynamic gestures to steer it or move it in forward and backward directions. Müezzinoğlu et al. developed a human-unmanned aerial vehicle interaction (UAV) interactive system using wearable gloves that could recognize 25 different hand gestures (Müezzinoğlu, 2021). The gloves contained flex sensors and IMUs and SVM yielded the best classification accuracy of 98.02%. In (Kundu, 2018), the authors used one IMU and two sEMG sensors on the hand to design a control mechanism for a wheelchair with 7 different hand gestures. A fusion of sEMG and eye-tracking based on visual information has been reported in (Zandigohar, 2021) to enhance the prediction of grasp intent. Data was recorded for 14 different grasp movements with 5 able-bodied subjects. Experimental results showed that the fusion of data from wearable and vision-based sensors improved the prediction of grasp intent by up to 13.66%.

Hand gesture and sign language recognition are being actively explored in the research community around the world. In this work, classification of both static and dynamic is performed using multiple wearable sensors on the dominant hand of the signer and a stacked deep learning architecture containing one-dimensional (1D) CNN and long short-term memory (LSTM) layers. The novel contributions of this work are

- Development multi-input stacked CNN-LSTM deep neural network for classification using multi-channel data
- Application of the multi-input CNN-LSTM model for classification of 50 commonly used signs from the Indian sign language
- Comparative analysis of the performance of the proposed multi-input model with conventional single-input approach in term of quantitative measures

The structure of the chapter is given as follows. The next section covers a review of deep learning architectures used with multi-channel and multi-modality data in various applications. Thereafter, details of the data acquisition device and the wearable sensor dataset for ISL used in this work are explained, followed by a detail explanation of the proposed multi-input stacked CNN-LSTM deep neural network for classification of the considered signs. Then experimental results on the ISL dataset are presented. The single-input model is tuned to optimize its performance. The multi-input models are designed with architecture same as that of the single-input model, but there is one model for each signal modality. The confusion matrices and the learning curves for the two models are also compared. Conclusions and scope of future work are listed at the end.

LITERATURE REVIEW

In this work, the focus is on the use of deep learning models for processing of multi-channel and multi-modality data for classification tasks. Some of the related work is briefly reviewed in Table 1. In certain research works, the multi-channel signals have been processed in a single CNN model as a multi-dimensional array. In (Park, 2020), for instance, signals from 4 sEMG sensors are collected from forearm and *imagified*, that is considered as an array with dimensions $4 \times$ the number of samples. The 2D sEMG array is given as input to a single CNN to achieve 100% accuracy for classification of 6 finger gestures from the Korean sign language. Using a 6-degreee of freedom (DoF) IMU, the authors of (Suri, 2019) stacked together the signals from 3-axis accelerometer and 3-axis gyroscope, along with 3-orientation time series data to give as input to a 1D Capsule Network. They achieved a classification accuracy of 94% for 20 sentences signed in ISL. A single-input CNN-LSTM model is used with multi-modal input for classification of 5 phases of gait cycle in (Kreuzer, 2021). The authors placed 11 sensors on legs and recorded accelerometer, gyroscope and barometer signals from all 11 sensors. The signals were stacked and given as input to a CNN-LSTM model for classification with up to 92% for unseen data. In (Dua, 2021), the authors designed a multi-head deep learning model, where each head comprised of CNN and Gated Recurrent Unit (GRU) layers, but with different filter sizes of 3, 7 and 11. The same raw sensor data was given as input to each model head and the feature maps generated by each head were concatenated an given to the final dense and softmax layers to classify the human activity with up to 97.21% accuracy.

Another approach is to process multi-channel or multi-modality data separately in different deep learning models and then combine the networks to form a single classifier. For example, the accelerometer, gyroscope and gravity signals from smartphone were processed in separate CNN-LSTM heads in (Yiming, 2021), and the resulting feature maps were concatenated before final classification layer. The authors report an F1-score of 0.9315 using the multi-head approach as compared to 0.9258 achieved using single-input ConvLSTM model for classification of human activity. A multibranch CNN-bidirectional LSTM (BiLSTM) model has been used on wearable sensor data to perform human activity recognition in (Challa, 2021). A multi-head Convolution Attention network is proposed for human activity recognition in (Zhang, 2020). The three-axis accelerometer data from smartphone was processed in three attention networks and concatenated for classification of six activities of daily living with 96.4% accuracy. In (Kraljević, 2020), when the RGB and depth data from video stream were processed in separate heads of a deep learning model and concatenated information was used classification, the average accuracy was 70.1% as compared to highest of 67.4% obtained without using sensor fusion. The authors employed 3D convolution layers (Conv3D) for extracting features from video streams in the two heads of the deep learning model. In (Gao, 2019), the authors processed the multi-modal inputs with RGB and depth information in two separate Inception-ResNet streams. Fusion of the class probabilities generated by the two streams was then used to predict the signs from ASL with 92.08% accuracy.

In this work, stacked CNN-LSTM model is employed for classification of ISL signs. As reported in (Gupta, Gupta, & Aswal, 2021), stacking of CNN with LSTM performs better as compared to CNN-only or LSTM-only model for classification of time series data, since the CNN model helps in extracting generic features and the LSTM layers capture the temporal dependencies in the data. Hence, stacked CNN-LSTM models have been employed by Kreuzer (2021) and Yiming (2021) for classification of time-series data. Moreover, multi-head processing is implemented with multi-modality wearable sensor data for classification of ISL signs, as explained in the next section.

Table 1. Review of sign language recognition systems

Reference	Data Acquisition	Algorithm	Best Results
(Park, 2020)	4-channel sEMG data	Single-input CNN model	100% for 6 signs from Korean sign language
(Suri, 2019)	One 6-DoF IMU	Single-input Capsule Network	94% for 20 ISL sentences
(Kreuzer, 2021)	11 6-DoF IMUs, also recording barometer signals	Single-input CNN-LSTM model	92% on unseen data for 5 phases of Gait
(Dua, 2021)	Accelerometer signals from smartphone available in WISDM Dataset	Multi-head CNN-GRU model	97.21% for human activity recognition
(Yiming, 2021)	Smartphone UCI data	Multi-head CNN-LSTM model	0.9315 F1-score human activity recognition
(Zhang, 2020)	3-axis accelerometer data	Multi-head Convolution Attention network	96.4% for recognition 6 human activities
(Kraljević, 2020)	RGB and depth data from video stream	Multi-head with Conv3D layers	70.1% for Croatian sign language
(Gao, 2019)	RGB and depth images	Two-stream CNN	92.08% for ASL

PROPOSED SIGN LANGUAGE RECOGNITION SYSTEM

Database of Indian Sign Language

The wearable sensor database for ISL signs, used in this work have been recorded by the author under the guidance of trained sign language interpreters. Five healthy subjects in the age group of 20-25 years, four of whom were right hand dominant and one was left-hand dominant, volunteered to participate in the study. The subjects were informed about the experiment protocol, which involved performing a sign when an audio was played and taking the hand back to rest, until the audio plays again and the sign is to be performed again. Each subject performed each sign 20 times, with a minimum rest duration of 3s in between consecutive repetitions. Fifty commonly used ISL signs were considered for processing. A variety of signs were considered, 31 of which such as 'Good', 'Fan' and numerals 0 to 9, were performed using only the dominant hand while the remaining 19 including signs for 'Lock', 'Add' and alphabets A, B, D required the use of both hands. Out of 50, 17 signs including the sign for 'Good' and the numerals 0 to 9, required the subject to hold certain hand posture and were static, whereas 33 signs such as those of 'Fan', 'Lock' and 'Add' required movement of hands and were dynamic in nature. The signs were recorded following the video dictionary made available by Ramakrishna Mission Vivekananda University (http://indiansignlanguage.org). Hence, 100 observations of each sign and 5000 total observations of the data were collected for processing.

For recording the wearable sensor dataset, three wireless sensors were placed on the dominant hand of the subject, as shown in Fig. 1, evenly spaced on the forearm. Three sensors were placed on the hand. Each sensor records 7 channels of time-series data, consisting of 1 channel sEMG, 3 channels accelerometer and 3 channels gyroscope data. The wireless sensors transmit the signals time-synchronously to the base station, which in turn sends the signals to the local computer for storage and display. Then the signals are processed offline for noise reduction and activity detection. Use of multi-modality sensors is know to enhance recognition capability (Sharma, 2019). Hence, the signals are required to be processed in an integrated manner for sign recognition. The deep learning models used in paper are explained in the next sub-section.

Figure 1. Data acquisition using wearable sensors

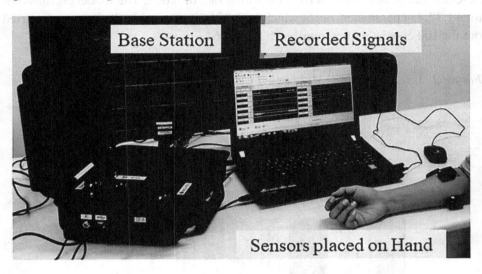

Mutli-Input Stacked CNN-LSTM Model

The time-series data is segmented to retain only the activity regions, while the rest durations are discarded. For uniform data size, the duration for all observations was limited to 2.7s, which is, in general, more than the time required to perform an isolated sign. The sEMG data is sampled at 1111.11 Hz and accelerometer and gyroscope data are recorded at 148.15 Hz. Hence, there are different number of samples in sEMG and IMU signals for the same activity duration. To train the single-head CNN-LSTM deep learning model, the signals were concatenated to form a single data matrix, as shown in Fig. 2a. The sEMG signals were resampled to 400 samples so that the number of samples are same as those in IMU data. The signals were divided into 10 time-steps, with 40 samples in each step. Since each sensor records 7 channels of data, three sensors yield an input data matrix with 21 features or channels. Generic features were extracted using two 1D convolutional layers with 6 and 12 filters, and kernel size 7. Sigmoid activation function was used in both the layers. Both convolutional layers were succeeded by pooling layers. Average pooling with window sizes 3 were utilized. Then, LSTM layers were included to extract time-dependencies in the data. The LSTM layers each having 100 units with drop out and recurrent dropout rate of 0.2 each, were followed by a batch normalization and dropout layers to avoid over-fitting and improve the speed of learning. The number of units in the LSTM layer were tuned as will be discussed in the results section. Finally, a dense layer with 50 units and softmax activation was applied to enable classification. The model was compiled using categorical cross-entropy loss function and Adam optimizer with learning rate of 0.002. The model consisted of total 1,32,454 parameters out of which 1,32,254 were trainable. Training was carried out for 100 epochs with a mini-batch size of 64.

Figure 2. Single-input and Multi-input CNN-LSTM architectures

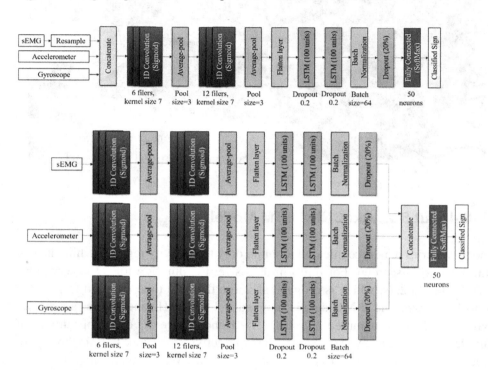

Another approach for processing multi-modal data is to design a multi-head deep learning model, having one head for each modality. As shown in Fig. 2b, the sEMG, accelerometer and gyroscope data were fed separately into the model with each of the heads. The sEMG data need not be resized, since it is not stacked with the IMU data, which was at a lower sample rate. The number of features available for sEMG were 3, while 9 channel accelerometer and gyroscope data were recorded, 3 channels from each sensor. Signals for all the modalities were divided into 10 time-steps with 300 and 40 samples of sEMG and IMU data, respectively. Other than the size of the input layers, the convolutional and LSTM layers are identical in configuration as explained in single-head CNN-LSTM architecture. In the multi-head architecture, the feature maps from the three modalities were concatenated prior to the final dense layer. The model consisted of total 5,34,698 parameters out of which 5,34,098 were trainable. The number of parameters in the multi-head model is almost 4 times than that in the single-head model. In the results section, the performance of the two model architectures is compared.

Table 2. Summary of single-input and multi-input models

	Single-input CNN-LSTM Model	Multi-input CNN-LSTM Model
Trainable parameters	1,32,254	5,34,098
Non-Trainable parameters	200	600
Total parameters	1,32,454	5,34,698

RESULTS AND DISCUSSION

The proposed multi-head stacked CNN-LSTM model was applied on the ISL database described above. The models were implemented in Python programming language using Keras and Tensorflow. First, the number of units in the LSTM layers of the single-input stacked CNN-LSTM model were varied to optimize the performance of the single-input model. Figure 3a and 3b show the loss and accuracy with respect to epochs for the single-input CNN-LSTM model. As the number of units in the two LSTM layers is increased beyond 150, the learning of the model slows down and the graphs converge after greater number of epochs. This is because of the increase in the parameters to be learned. Moreover, there is no significant improvement in the performance of the model in term of the loss or the accuracy.

Figure 3. Tuning of LSTM layers in single-input stacked CNN-LSTM model (a) Test loss (b) Test accuracy

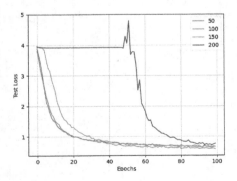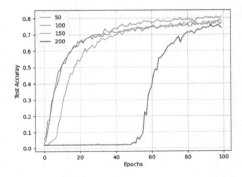

As the number of units increase from 50 to 100, the classification accuracy at the end of 100 epochs increases from 77.1% to 81.2%, respectively. When the units are further increased in number, the classification accuracy degrades to 77.9% and 73.9% for 150 and 200 units, respectively. The lowest value of loss and the highest accuracy with test data are achieved with 100 units in each LSTM layer. Hence, this configuration is retained for the remaining analysis.

Next, the performance of the single-input approach is compared with that of the multi-input approach in terms of various quantitative criteria. Figure 4 shows the training and test loss with respect to epochs for single- and multi-input CNN-LSTM models. It is evident that both the models are converging as expected, with no indication of overfitting. The multi-input approach performs better as compared to the single-input approach in terms of loss value. Similarly, the multi-input approach also outperforms the single-input approach in terms of accuracies for both training and testing data, as seen in Figure 5. The multi-input CNN-LSTM model yield significantly higher accuracy as compared to the single-input counterpart. While the classification accuracy at the end of 100 epochs was 84.3% on training data and 81.2% on test data for the single-input model, it was 89.9% and 86.1% for training and test data, when multi-input model was used.

Figure 4. Comparison of Loss curves (a) Training loss (b) Test loss

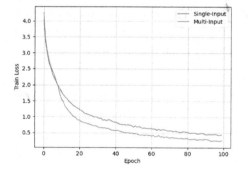 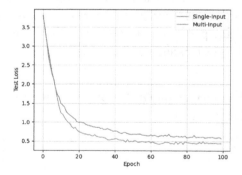

Figure 5. Comparison of Accuracy curves (a) Training accuracy (b) Test accuracy

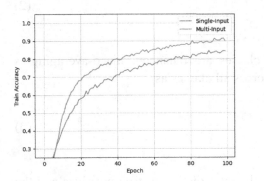 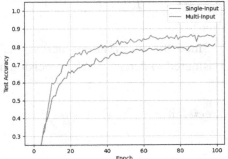

Confusion matrices for all 1500 observations tested using the single- and multi-input CNN-LSTM models are shown in Figure 6 and Figure 7, respectively. There are 30 observations of each sign available in the test data. Both the models seem to perform better on the dynamic signs involving motion of hands, as compared to the static signs. For instance, signs 21 to 32 represent static signs from numerals 1 to 9 and three alphabets A, B and C. These signs only require a slight change in configuration of fingers and are difficult to predict, particularly because the sensors measuring the muscle activity are placed on the forearm. Also, certain signs are distinguished by the change in the orientation of the non-dominant hand. For example, sign 46 for work and sign 47 for meat are almost identically performed by the dominant hand. Hence, there is more confusion between these two signs because the sensors are only placed on the dominant hand. However, the multi-input model improves the overall classification of these signs. Further improvement may require the use of more sensors or a different sensing modality.

Finally, the performance of the two approaches is also compared in terms of other performance metric, as given in Table 3. The classification accuracy may be calculated from the confusion matrix by taking the ratio of the number of correct predictions, that is the sum of diagonal elements and the total number predictions made by the model, as

$$Accuracy = \frac{Total\ number\ of\ correct\ predictions}{Total\ number\ of\ predictions}. \tag{1}$$

Precision indicates the number of true positives within the number of predicted positives, which includes the true positives TP_i and false positives FP_i for the class $i, i = 1, \ldots N$, where N is the total number of classes (Grandini, 2020),

$$Precision_i = \frac{TP_i}{TP_i + FP_i}. \tag{2}$$

Since the observations across the classes are balanced, maco-averaging is considered to determine the precision across all classes as,

$$Precision = \frac{1}{N} \sum_{i=1}^{N} Precision_i. \tag{3}$$

As given in (4), recall for class i is the ratio of sum of true positives and the sum of true positives and false negatives for that class,

$$Recall_i = \frac{TP_i}{TP_i + FN_i}, \tag{4}$$

and recall across all classes is determined using macro-averaging as (Grandini, 2020),

$$Recall = \frac{1}{N} \sum_{i=1}^{N} Recall_i. \tag{5}$$

Macro F1-score is found as the weighted mean of precision and recall, given by

$$F1 = 2 * \frac{Precision * Recall}{Precision + Recall}. \tag{6}$$

As observed from Table 3, overall, the performance of multi-input model is much better than that of the single-input model in terms of precision, recall, F1-score and average accuracy across all signs. There is a 5% improvement observed in precision, recall, F1-score as well as classification accuracy of the proposed multi-input CNN-LSTM model as compared to the single-input model.

Figure 6. Confusion matrix for single-input CNN-LSTM

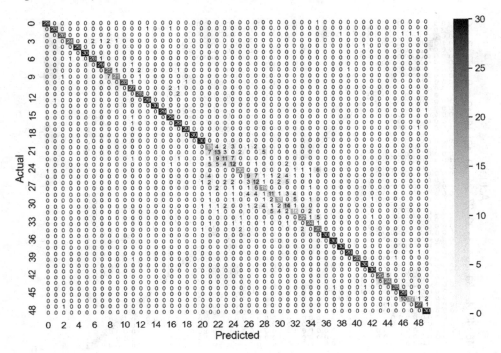

This study shows that although the sampling rates of sensing modalities may be the same and their signals may be stacked together to be given as input to a deep learning model, however, they should be processed in different deep learning models for feature extraction because they convey different information. The feature maps learned on individual sensing modalities when concatenated and used for classification performed better than the feature map extracted from the raw signals stacked together. In future, work will be carried out towards improving the classification accuracy by processing signals captured from both hands. The multi-input approach will need to be designed for a two-hand recording. The models could also be compressed so that the light-weight models may be deployable and suitable for development of a practical wearable system for sign language recognition.

Figure 7. Confusion matrix for multi-input CNN-LSTM

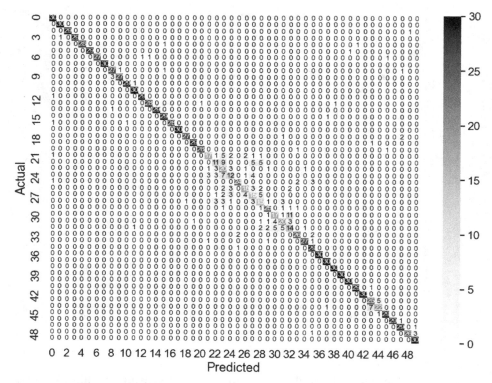

Table 3. Comparison of performance of single-input and multi-input models

	Single-input CNN-LSTM Model	**Multi-input CNN-LSTM Model**
Precision	81.5%	86.4%
Recall	81.2%	86.1%
F1-score	0.81	0.86
Accuracy	81.2%	86.1%

CONCLUSION

In this work, the use of wearable sensors and deep learning for sign language recognition is explored. Sign language recognition has been carried out using multiple wireless sensors that record 7 channel data, 1 surface electromyogram, and

6-channel IMU data. Initially, the signals are stacked together and given as input to a single-head stacked CNN-LSTM model. Results are also presented for tuning the model by varying the number of units in the LSTM layers. An average classification accuracy of 81.2% is obtained for 50 ISL signs with single-head stacked CNN-LSTM model. Use of single-head deep learning model poses limitations for processing multi-modality input data such as the signals may have to be resampled to match the dimensions and the features are extracted in a similar manner from all signals, although they have very distinct characteristics. A better approach is to process the signals from different modalities, such as sEMG, accelerometer and gyroscope, using separate deep learning models. Hence, three stacked CNN-LSTM models are used to extract feature maps from each type of signal. Then, the feature maps are concatenated before applying the softmax activation to perform classification. Using the multi-head stacked CNN-LSTM model, the average classification accuracy improved to 86.1%. The performance of the considered models is also compared in terms of confusion matrix, precision, recall and F1 score. Overall, the multi-head approach outperforms the conventional single-input model. The limitation of the proposed approach is the 4-fold increase in the number of trainable parameters. In future, work needs to be carried out to develop model compression algorithms that would yield comparable accuracies with relatively smaller number of parameters.

ACKNOWLEDGMENT

This research was supported by the Science and Engineering Research Board, a statutory body from the Department of Science and Technology, Government of India. [grant number ECR/2016/000637].

REFERENCES

Banjarey, K., Sahu, S. P., & Dewangan, D. K. (2021, April). A Survey on Human Activity Recognition using Sensors and Deep Learning Methods. In *2021 5th International Conference on Computing Methodologies and Communication (ICCMC)* (pp. 1610-1617). IEEE. 10.1109/ICCMC51019.2021.9418255

Challa, S. K., Kumar, A., & Semwal, V. B. (2021). A multibranch CNN-BiLSTM model for human activity recognition using wearable sensor data. *The Visual Computer*, 1–15. doi:10.100700371-021-02283-3

Dai, D., Zhuang, W., Shen, Y., Li, L., & Wang, H. (2020, July). Design of Intelligent Mobile Robot Control System Based on Gesture Recognition. In *International Conference on Artificial Intelligence and Security* (pp. 101-111). Springer. 10.1007/978-981-15-8086-4_10

Dong, Y., Liu, J., & Yan, W. (2021). Dynamic Hand Gesture Recognition Based on Signals From Specialized Data Glove and Deep Learning Algorithms. *IEEE Transactions on Instrumentation and Measurement, 70*, 1–14. doi:10.1109/TIM.2021.3077967

Dua, N., Singh, S. N., & Semwal, V. B. (2021). Multi-input CNN-GRU based human activity recognition using wearable sensors. *Computing, 103*(7), 1–18. doi:10.100700607-021-00928-8

Gałka, J., Mąsior, M., Zaborski, M., & Barczewska, K. (2016). Inertial motion sensing glove for sign language gesture acquisition and recognition. *IEEE Sensors Journal, 16*(16), 6310–6316. doi:10.1109/JSEN.2016.2583542

Gao, Q., Ogenyi, U. E., Liu, J., Ju, Z., & Liu, H. (2019, September). A two-stream CNN framework for American sign language recognition based on multimodal data fusion. In *UK Workshop on Computational Intelligence* (pp. 107-118). Springer.

Grandini, M., Bagli, E., & Visani, G. (2020). *Metrics for multi-class classification: an overview.* arXiv preprint arXiv:2008.05756.

Gupta, R., Gupta, A., & Aswal, R. (2021, January). Time-CNN and Stacked LSTM for Posture Classification. In *2021 International Conference on Computer Communication and Informatics (ICCCI)* (pp. 1-5). IEEE. 10.1109/ICCCI50826.2021.9402657

Gupta, R., & Kumar, A. (2021). Indian sign language recognition using wearable sensors and multi-label classification. *Computers & Electrical Engineering, 90*, 106898. doi:10.1016/j.compeleceng.2020.106898

Khomami, S. A., & Shamekhi, S. (2021). Persian sign language recognition using IMU and surface EMG sensors. *Measurement, 168*, 108471. doi:10.1016/j.measurement.2020.108471

Kolivand, H., Joudaki, S., Sunar, M. S., & Tully, D. (2021). A new framework for sign language alphabet hand posture recognition using geometrical features through artificial neural network (part 1). *Neural Computing & Applications, 33*(10), 4945–4963. doi:10.100700521-020-05279-7

Kraljević, L., Russo, M., Pauković, M., & Šarić, M. (2020). A Dynamic Gesture Recognition Interface for Smart Home Control based on Croatian Sign Language. *Applied Sciences (Basel, Switzerland), 10*(7), 2300. doi:10.3390/app10072300

Kreuzer, D., & Munz, M. (2021). Deep Convolutional and LSTM Networks on Multi-Channel Time Series Data for Gait Phase Recognition. *Sensors (Basel)*, *21*(3), 789. doi:10.339021030789 PMID:33503947

Kudrinko, K., Flavin, E., Zhu, X., & Li, Q. (2020). Wearable Sensor-Based Sign Language Recognition: A Comprehensive Review. *IEEE Reviews in Biomedical Engineering*, *14*, 82–97. doi:10.1109/RBME.2020.3019769 PMID:32845843

Kundu, A. S., Mazumder, O., Lenka, P. K., & Bhaumik, S. (2018). Hand gesture recognition based omnidirectional wheelchair control using IMU and EMG sensors. *Journal of Intelligent & Robotic Systems*, *91*(3), 529–541. doi:10.100710846-017-0725-0

Masood, S., Srivastava, A., Thuwal, H. C., & Ahmad, M. (2018). Real-time sign language gesture (word) recognition from video sequences using CNN and RNN. In *Intelligent Engineering Informatics* (pp. 623–632). Springer. doi:10.1007/978-981-10-7566-7_63

Muezzinoglu, T., & Karakose, M. (2021). An Intelligent Human–Unmanned Aerial Vehicle Interaction Approach in Real Time Based on Machine Learning Using Wearable Gloves. *Sensors (Basel)*, *21*(5), 1766. doi:10.339021051766 PMID:33806388

Park, J. J., & Kwon, C. K. (2021). Korean Finger Number Gesture Recognition Based on CNN Using Surface Electromyography Signals. *Journal of Electrical Engineering & Technology*, *16*(1), 591–598. doi:10.100742835-020-00587-3

Ramakrishna mission Vivekananda University, Coimbatore Campus. (2021). *Indian Sign Language Dictionary*. http://indiansignlanguage.org

Rastgoo, R., Kiani, K., & Escalera, S. (2020). Sign language recognition: A deep survey. *Expert Systems with Applications*, 113794.

Sharma, S., Gupta, R., & Kumar, A. (2019, March). On the use of multi-modal sensing in sign language classification. In *2019 6th International Conference on Signal Processing and Integrated Networks (SPIN)* (pp. 495-500). IEEE. 10.1109/SPIN.2019.8711702

Suri, K., & Gupta, R. (2019). Continuous sign language recognition from wearable IMUs using deep capsule networks and game theory. *Computers & Electrical Engineering*, *78*, 493–503. doi:10.1016/j.compeleceng.2019.08.006

Wadhawan, A., & Kumar, P. (2020). Deep learning-based sign language recognition system for static signs. *Neural Computing & Applications*, *32*(12), 7957–7968. doi:10.100700521-019-04691-y

Wadhawan, A., & Kumar, P. (2021). Sign language recognition systems: A decade systematic literature review. *Archives of Computational Methods in Engineering, 28*(3), 785–813. doi:10.100711831-019-09384-2

Yiming, Z., & Liuai, W. (2021). Human Motion State Recognition Based on Multi-input ConvLSTM. *Journal of Physics: Conference Series, 1748*(6), 062075. doi:10.1088/1742-6596/1748/6/062075

Yuan, G., Liu, X., Yan, Q., Qiao, S., Wang, Z., & Yuan, L. (2020). Hand gesture recognition using deep feature fusion network based on wearable sensors. *IEEE Sensors Journal, 21*(1), 539–547. doi:10.1109/JSEN.2020.3014276

Zandigohar, M., Han, M., Sharif, M., Gunay, S. Y., Furmanek, M. P., Yarossi, M., . . . Schirner, G. (2021). *Multimodal Fusion of EMG and Vision for Human Grasp Intent Inference in Prosthetic Hand Control.* arXiv preprint arXiv:2104.03893.

Zhang, H., Xiao, Z., Wang, J., Li, F., & Szczerbicki, E. (2019). A novel IoT-perceptive human activity recognition (HAR) approach using multihead convolutional attention. *IEEE Internet of Things Journal, 7*(2), 1072–1080. doi:10.1109/JIOT.2019.2949715

KEY TERMS AND DEFINITIONS

Accelerometer: A sensor that can record acceleration due to gravity as well as linear acceleration caused by motion of the sensor.

Classification Accuracy: The number of times a classification models makes a correct prediction as compared to the total number of predictions made by the classifier, stated in terms of percentage.

Convolutional Neural Network: Convolutional neural network, also known as ConvNet or CNN is a deep feed forward neural network that makes use of operation of convolution with sliding kernel to generate representative feature map of the input.

Electromyogram: A recording of the muscle potentials, used to detect activity or in monitoring health of muscle.

Gyroscope: A sensor that can record turn rate in terms of angle per unit time.

Long Short-Term Memory: Long short term memory or LSTM is a recurrent neural network that contains feedback connections and are commonly used with time-series data.

Wearable Sensors: Worn on body and consisting of a sensor that can record physical phenomenon in terms of electrical signal.

Chapter 9
Lightweight ConvNet Model for American Sign Language Hand Gesture Recognition

Shamik Tiwari

https://orcid.org/0000-0002-5987-7101
University of Petroleum and Energy Studies, India

ABSTRACT

Deaf and hard-of-hearing persons practice sign language to converse with one other and with others in their community. Even though innovative and reachable technology is evolving to assist persons with hearing impairments, there is more scope of effort to be achieved. Computer vision applications with machine learning procedures could benefit such persons even more by allowing them to converse more effectively. That is precisely what this chapter attempts to do. The authors have suggested a MobileConvNet model that could recognise hand gestures in American Sign Language. MobileConvNet is a streamlined architecture that constructs lightweight deep convolutional neural networks using depthwise separable convolutions and provides an efficient model for mobile and embedded vision applications. The difficulties and limitations of sign language recognition are also discussed. Overall, it is intended that the chapter will give readers a thorough overview of the topic of sign language recognition as well as aid future research in this area.

INTRODUCTION

Day by day, computing gadgets are becoming an increasingly important aspect of our life. As the need for such computing devices grew, so did the need for simple and

DOI: 10.4018/978-1-7998-9434-6.ch009

effective computer interfaces. As a result, systems that use vision-based interaction and control are becoming more widespread, and as a result, gesture recognition is becoming increasingly popular in the research community due to a variety of reasons (Ghanem et al., 2017). Hand gestures are a type of body language that can be communicated by the position of the fingers, the centre of the palm, and the shape formed by the hand. Static and dynamic hand movements can be distinguished. The static gesture refers to the hand's fixed shape, whereas the dynamic gesture is made up of a sequence of hand movements such as waving. Gesture interaction is a well-known technology that can be utilized in a variety of applications, containing sign language translation, sports, human-robot interaction, and human-machine interaction in general. Hand-gesture recognition systems are also used in medical applications, where bioelectrical signals are used instead of eyesight to identify gestures. Gestures can be categorized broadly into the following groups (Banjarey et al., 2021; Tiwari, 2018a).

- Head and Face Gestures: moving the head, eye movement direction, lifting the eyebrows, blinking, curling the nostrils, raising the mouth to speak, smiles, delight, contempt, panic, hate, grief, disdain etc. are examples of head and face gestures.
- Hand and Arm Gestures: Recognition of hand positions, sign languages, and entertainment and gaming applications are all possible using hand and arm gestures.
- Body Gestures: Full-body motion is involved in body gestures, such as following the motions of two individuals engaging together, evaluating a dancer's movements to generate corresponding music and graphics, and detecting human postures for medical rehabilitation and physical education (Challa et al., 2021).

There are two sorts of sensors applied to recognise hand gestures referred as contact sensors and non-contact sensors. Contact approaches analyse the signal obtained from contact sensors bound to the wrist or arm, to identify gestures (Bantupalli & Xie, 2018; Dua et al., 2021). Contact-type devices take longer time to measure since they must touch and then traverse the item. They have a superior identification range than non-contact systems, as they are not restricted by range or sensor sight, and they can obtain relatively precise information owing to direct touch. According to numerous research, non-contact methods are mostly established with machine vision equipment such as leap motion controller, camera sensors, Kinect. These sensors do not attached to the human body. Non-contact sensors are much less susceptible to sensor wear and will not diminish a target's motion. The rest of the chapter is divided into literature review, material and methods, experiment and results, and conclusion respectively in sections 2, 3, 4 and 5.

LITERATURE REVIEW

This section analyses the available literature on gesture recognition systems for HCI by classifying it according to certain essential features. It also examines the advancements that are required to improve current hand gesture detection systems in order for them to be extensively employed for optimal HCI in the future.

To assist conversation among signers and non-signers, Bantupalli et al. (Bantupalli & Xie, 2018) have designed a machine vision system that provides sign language interpretation to text. This suggested application extracts temporal and spatial characteristics from video sequences. Then, for identifying spatial features, utilise Inception, a CNN model. Then, to train on temporal characteristics, utilise an RNN. The American Sign Language Dataset was used in this study. Garcia and Viesca (Garcia & Viesca, 2016) have demonstrated the creation and deployment of a CNN-based American Sign Language fingerspelling translator. To apply transfer learning, they use a pre-trained GoogLeNet structure trained on the ILSVRC2012 dataset as well as the Surrey University and Massey University ASL datasets. They have developed a robust model that correctly recognises letters a-e in the majority of instances and accurately identifies letters a-k in the majority of instances with first-time users.

Dong et al. (Dong et al., 2015) have developed a strategy for ASL identification employing a Kinect depth camera from Microsoft. A feature extraction method based on depth contrast per-pixel categorization is used to create a segmented hand configuration. Then, given kinematic restrictions, a hierarchical mode-seeking approach is designed to localise hand joint positions. In conclusion, a RF classifier is developed to recognise signs gestures by the use of joint angles. A publicly accessible sign image dataset from Surrey University is used to assess the method's performance. The approach was able to recognise 24 static ASL letter signs with an accuracy of above 90%. Jin (Jin et al., 2016) have used image processing approaches to identify the images of sign language. Canny edge detection and seeded region growing algorithms are used to separate the hand gesture from background. SURF is then applied for feature extraction. Following that, SVM is applied to categorise a dataset of gesture images, with the trained dataset being utilised to recognise future sign gesture. The suggested system has been developed and tested on android smartphone platforms. The experimental findings indicate that this system is able to recognise and translate 16 different ASL motions with a 97.13 percent overall accuracy.

McKee et al. (McKee et al., 2015) have created a method for translating, adapting, and developing an affordable medical competence instrument in ASL, as well as a cross-sectional strategy to analyze the extent and relates of poor fitness literacy among deaf ASL practitioners and hearing English talkers. The proposed approach, which is accessible on a self-administered system, has shown a strong link to reading literacy. The high frequency of Deaf ASL users with low health literacy necessitates additional interventions and study. A system that recognises sign language in real time is developed by Taskiran et al. (Taskiran et al., 2018). In this work, the ConvNet is trained using the ASL dataset, and the achieved test accuracy is 100 percent. The model parameters and weights are maintained and utilized for the real-time scenario after completion of the training. The skin colour for a definite frame of hand gesture has been calculated in the real-time system, and the hand motion is decided using the convex hull technique, and the hand motion is established in real-time using the recorded model with corresponding parameters and weights. The system has achieved 98.05 percent accuracy in real time.

Ameen and Vadera (Ameen & Vadera, 2017) have investigated the use of deep learning in the interpretation of sign language and developed a ConvNet to categorise finger spelling images with both the depth data and image intensity. The constructed model is put to the test by using it to solve the challenge of ASL finger spelling recognition. The generated model outperforms prior tests, with precision of 82 percent and sensitivity of 80 percent, according to the evaluation. The evaluation's confusion matrix indicates the underlying issues with categorising several specific indicators, which are explored in the study. Lee et al. (Lee et al., 2021) have suggested a prototype for an ASL learning. A whack-a-mole game with a real-time sign identification system would be the application. Because ASL alphabets contain both dynamic and static signs (J, Z), the classification method used is the LSTM RNN with KNN approach, which is based on the processing of input sequences. The angles between the fingers, radius of the sphere and the distance between the placements of the fingers are all retrieved as features. A total of 2600 examples were used to train the model, with 100 examples for each alphabet. The results of the experiments showed that the classification performance for 26 ASL alphabets is 99.44 percent accurate on average and 91.82 percent with the usage of a leap motion controller.

Noroozi et al. (Noroozi et al., 2018) have presented emotive body gestures as a component and discuss broad issues such as gender disparities and cultural sensitivity. They offer person detection with dynamic and static body posture assessment approaches in 3D and RGB colour images, as well as a system for emotional body gesture identification. They also examine multi-modal approaches to emotion recognition that integrate speech or facial with body motions for gesture identification. To deal with the temporal component is the most important difficulty in human gesture detection. In the convolutional layer, the some authors use 3D filters. Some notable work in this group are presented by Baccouche et al. (Baccouche et al., 2011). Baccouche et al. offered a fully automated deep learning model that learns to characterise human activities without any former knowledge. Initial stage of this schemes is based on the extension of 2D CNN to 3D CNN, learns spatial-temporal properties automatically. The learned features are then used to train a RNN to categorize each sequence based on their temporal evolution for each timestamp. Deep neural networks have recently outpaced prior state-of-the-art machine learning approaches in a variety of fields, particularly for machine vision applications. The following are some of the most important deep learning models for computer vision problems. CNN (Tiwari, 2018b), Deep Boltzmann Machine, DBN, Capsule networks (Tiwari & Jain, 2021), Auto Encoder, GAN (Gui et al., 2020), and RNN including LSTM and GRU. In terms of vision-based approaches, future research must address key issues such as hand gesture classification accuracies, articulated hand pose estimation accuracy, and computational efficiency. Most present vision-based systems can only identify static motions, hence new ways for managing dynamic gestures are required.

MATERIAL AND METHODS

This section provides details of CNN, MobileNet, proposed methodology and description of sign language dataset.

Convolution Neural Network

The ConvNet is a type of deep neural network model designed for working with 2-D image data, though it can also be used with 1-D and 3-D image data. Speech recognition, pattern classification, and text processing are some of the other applications that use ConvNets. MLP were utilised to develop image classifiers before convolutional neural networks. The convolutional layer, which gives the network its name, is at the heart of the convolutional neural network. This layer conducts an action called a 'convolution' (Sharma & Tiwari, 2021; Tiwari et al., 2021). Multiple convolutional layers are placed on top of each other in convolutional neural networks, each competent of identifying more complex shapes. For example, handwritten digits can be recognised with 3 or 4 convolutional layers, while human faces can be distinguished with 20 layers. The input image is transformed using a convolution layer in order to extract features from it. The image is convolved with a filter (or kernel) in this transformation. A kernel is a small matrix that is smaller in height and width than the image to be convolved. A representative diagram of convolution layer is given in Figure 1.

Figure 1. Demonstration of convolution operation between input image and convolution filter producing output map

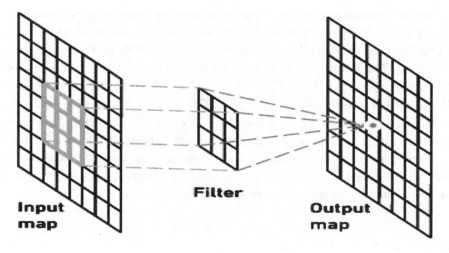

Input map

Filter

Output map

Other important layer of ConvNet models are pooling layer, activation layer and dense layer. Using pooling layers, the dimensions of the feature maps are lowered. As a result, both the number of parameters to train and the network's processing are lowered. The pooling layer sums together the features present in a region of the feature map obtained by a convolution layer. The activation layer's aim is to establish non-linearity into a neuron's output. An activation function is a function that is introduced to a ConvNet to assist it in learning complex patterns from data. In a convolution layer, the ReLu or softmax function is commonly utilised as an activation function. The fully connected layer is a deep-connected neural network layer, meaning that each neuron in the dense layer accepts input from all neurons in the earlier layer. This layer is used to identify images based on convolutional layer output.

Lightweight Network Structure-MobileNet

The MobileNet model, as its name implies, is intended for usage in mobile applications and is TensorFlow's first mobile computer vision model. MobileNets are low-latency, low-power models that have been parameterized to match the resource restrictions of various use cases. Identification, localization, classification, embeddings, and segmentation can all be constructed on top of them (Srinivasu et al., 2021). Google's MobileNet series network is a concept for mobile and embedded devices that aims to strike a balance between accuracy and speed of operation. MobileNet is a streamlined architecture that constructs lightweight deep convolutional neural networks using depthwise separable convolutions and provides an efficient model for mobile and embedded vision applications. A MobileNet contains 28 layers if depthwise and pointwise convolutions are counted separately. The width multiplier hyperparameter can be adjusted to decrease the number of parameters in a conventional MobileNet to 4.2 million. Depth separable convolution is a decomposition of ordinary convolution into $k * k$ depth-wise (DW) and $1x1$ point-wise components. For each input channel, deep convolution utilises a single convolution kernel, and point-wise convolution combines the deep convolution outputs. The above two things are done via a typical convolution. The depth separable convolution, on the other hand, splits them into two steps. The computational complexity and size of the model are greatly reduced as a result of this deconstruction (Chen et al., 2021).

$$Complexity = \frac{Depth\text{-}wise\ separable\ convolution}{Standard\ convolution} = \frac{1}{K^2} + \frac{1}{C_{out}} \approx \frac{1}{K^2} \qquad (1)$$

Figure 2. Comparison of standard convolution and depth separable convolution

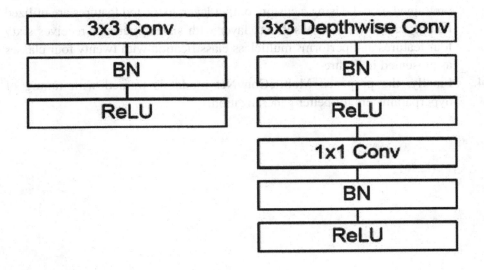

Proposed MobileConvNet Model for Sign Language Gesture Recognition

The proposed model has utilized transfer learning approach. One of the most often used approaches for classification and recognition is transfer learning. It is the transmission of information or weights gained from problem of one domain to solve other related problems same domain. The pre-trained MobileNet is utilized to design the proposed model of sign language gesture recognition. The pre-trained MobileNetV2 was created with a categorization capacity of 1000 classes in consideration. The output layer has been replaced by a global pooling layer. One advantage of global average pooling over dense layers is that it enforces interactions between feature maps and classes, making it more native to the convolution structure. The following are the procedures for building the proposed model.

1. MobilNetV2 receives the input images and generates the initial feature maps. Except that it employs inverted residual blocks with bottlenecking features, MobileNetV2 is quite identical to the original MobileNet. It has a significantly fewer number of parameters than the original MobileNet. Any image size bigger than 32×32 is supported by MobileNets.
2. Global pooling block transforms the output feature map into a 1-D vector consisting of sixty four features.
3. The output layer, which works as a feature extractor, is excluded from the pretrained model. Using a custom output layer, extracted features are utilized for the categorization. A softmax layer with sixteen neurons receives sixty four features and performs multiclass classification with twenty four classes as presented in Figure 5.
4. Finally, the proposed MobileConvNet model is trained with tuning of hyperparameters for better generalization.

Figure 3. Proposed lightweight MobileConvNet model for sign language gesture recognition

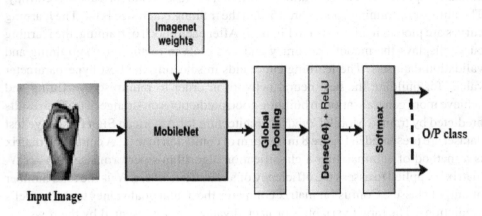

Sign Language Image Dataset

The MNIST image dataset of handwritten digits is a well-recognized image dataset for image-based machine learning algorithms, but academics are working to improvise it and produce drop-in alternates that are further difficult for classifiers. The Sign Language MNIST is offered here to encourage the computer vision research groups to produce more drop-in alternatives. It uses the same CSV format as the other MNISTs, with labels and pixel values in single rows. This dataset consists of sign language with 24 classes of letters, where J and Z are excluded since these alphabet need motion (tecperson, 2017). Total number of images are 27456 and 7172 respectively for training and testing samples. The sample images are provided in Figure 4 and individual count of the images for each class is given in Figure 5. The size of the original images was 28X28 pixels. However, these images are resized to 32X32 as minimum size required for a MobileNet architecture.

EXPERIMENT AND RESULTS

MobileConvNet is built with Keras, an open source deep learning framework. MobileConvNet can be trained from start to finish, with the parameters updated via stochastic gradient descent. We used MobileNetv2's pre-trained weight on ImageNet to initialise parameters before training because MobileConvNet shares a part of the MobileNetv2 network topology. The following are the hyperparameter values in the model: the model's initial learning rate is 0.01. The learning rate is dropped by 0.5

times per 10 epochs. The weight and momentum degradation rates are 0.9 and 0.0005, respectively. The activation is softmax, the loss is categorical cross entropy. The number of training epochs are 15, and the training batch size is 64. The learning curves are plotted in Figure 6 and Figure 7. After each epoch of training, the learning curve displays the model's accuracy and loss plots in relation to the training and validation data sets. The learning curve aids in selecting the best hyperparameter values for building the best neural network in order to minimise overfitting and achieve more generalisation in building good predictions on strange data. The results predicted by trained MobileConvNet architecture for American Sign Language test dataset are presented in Figure 8 in the form of confusion matrix. A confusion matrix is a method of summarising a classification algorithm's performance. A $N \times N$ matrix is applied to assess the efficiency of a classifier, where N is the total number of output classes. Confusion matrix compares the actual goal values to the model's predictions. The target variable's projected values are represented by the rows.

Figure 4. Sample images from American Sign Language image dataset

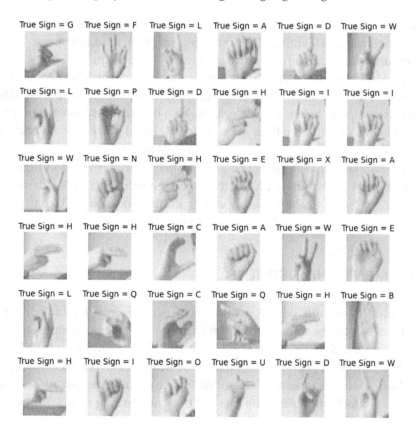

Figure 5. Count of the sign language gesture images for each class

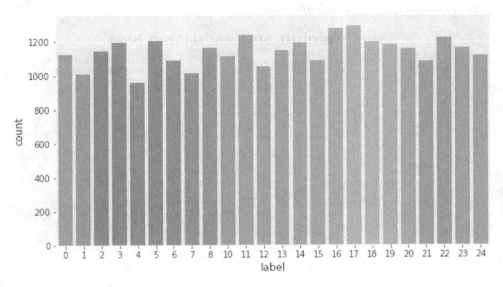

Figure 6. Training and validation accuracy curves of MobileConvNet classifier for American Sign Language classification

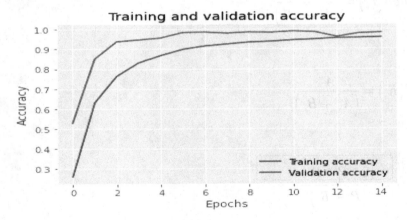

Figure 7. Training and validation loss curves of MobileConvNet classifier for American Sign Language classification

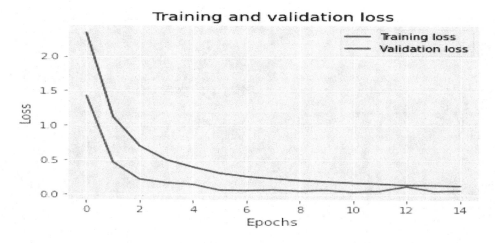

Further, precision, recall, and F-score are calculated to assess the performance of the MobileConvNet model. These metrics are computed as follows (Juba & Le, 2019; Powers, 2020).

$$Precision\left(P_r\right) = \frac{A^+}{\left(A^+ + B^+\right)} \tag{2}$$

$$Recall\left(R_c\right) = \frac{A^+}{\left(A^+ + B^-\right)} \tag{3}$$

$$F\text{-}Score = \frac{\left(2 * P_r * R_c\right)}{\left(P_r + R_c\right)} \tag{4}$$

$$Accuracy = \frac{\left(A^+ + B^-\right)}{\left(A^+ + A^- + B^- + B^+\right)} \tag{5}$$

Figure 8. Results in terms of confusion matrix predicted by MobileConvNet model for American Sign Language test dataset

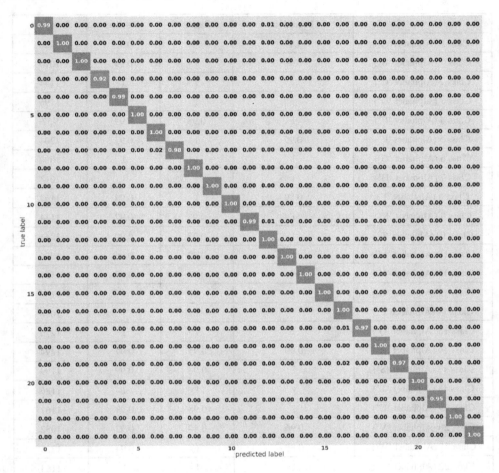

Here A[+], B[+], A[-], and B[-] are the true positive, false positive, true negative, and false negative classified gestures by the MobileConvNet model, respectively. These metrics are offered in Table 1. It can be noticed from the results that precision, recall and F-score are 1 for alphabets C, D, G, H, O, P, Q, T and X. The overall classification accuracy of the model is 98% and adequate values of other performance metrics as provided in the table. A useful method for evaluating diagnostic tests and predictive models is ROC analysis. It can be used to evaluate accuracy between tests or predictive models or to assess accuracy quantitatively (Bowers & Zhou, 2019). ROC curves are plotted in Figure 9. From these ROC curves, it can be observed that area under curve is 1 for all classes that confirms the robustness of the MobileConvNet model.

Table 1. Results using lightweight MobileConvNet model for American Sign Language classification

	Precision	Recall	F-score	Support
Class 0 (Alphabet 'A')	0.97	0.88	0.93	1126
Class 1 (Alphabet 'B')	0.98	1	0.99	1010
Class 2 (Alphabet 'C')	1	1	1	1144
Class 3 (Alphabet 'D')	1	1	1	1196
Class 4 (Alphabet 'E')	1	0.98	0.99	957
Class 5 (Alphabet 'F')	0.95	1	0.97	1204
Class 6 (Alphabet 'G')	1	1	1	1090
Class 7 (Alphabet 'H')	1	1	1	1013
Class 8 (Alphabet 'I')	0.89	1	0.94	1162
Class 9 (Alphabet 'K')	1	0.93	0.97	1114
Class 10 (Alphabet 'L')	1	0.97	0.99	1241
Class 11 (Alphabet 'M')	1	0.92	0.96	1055
Class 12 (Alphabet 'N')	0.96	0.99	0.97	1151
Class 13 (Alphabet 'O')	1	1	1	1196
Class 14 (Alphabet 'P')	1	1	1	1088
Class 15 (Alphabet 'Q')	1	1	1	1279
Class 16 (Alphabet 'R')	0.99	0.95	0.97	1294
Class 17 (Alphabet 'S')	1	0.98	0.99	1199
Class 18 (Alphabet 'T')	1	1	1	1186
Class 19 (Alphabet 'U')	1	0.98	0.99	1161
Class 20 (Alphabet 'V')	0.96	0.89	0.92	1082
Class 21 (Alphabet 'W')	0.89	1	0.94	1225
Class 22 (Alphabet 'X')	1	1	1	1164
Class 23 (Alphabet 'Y')	0.95	1	0.97	1118
Accuracy 0.98				

Figure 9. ROC curves for sign language classification

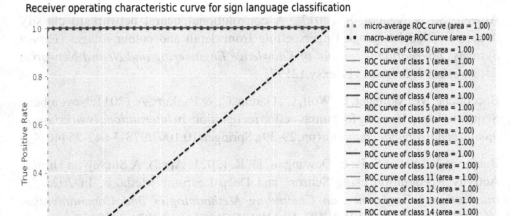

Receiver operating characteristic curve for sign language classification

CONCLUSION

Sign languages are used by disabled individuals all over the world to communicate. New technology innovations, such as cell phones, provide customers with a slew of new features. If such mobile devices can recognise sign languages, sign language users will be able to access far more user-friendly smartphone applications. This chapter proposes a lightweight convolution neural network to recognize the American Sign Language. Such a system can be utilized on smartphone for gesture recognition. Despite the fact that American Sign Language has earned widespread acceptance in the United States, handful ASL applications are being developed for academic purposes but real-time sign recognition technologies are still missing. It is required to enable the creation of a learning application with a real-time sign gesture recognition that aims to increase ASL learning efficiency in future.

REFERENCES

Ameen, S., & Vadera, S. (2017). A convolutional neural network to classify American Sign Language fingerspelling from depth and colour images. *Expert Systems: International Journal of Knowledge Engineering and Neural Networks*, *34*(3), e12197. doi:10.1111/exsy.12197

Baccouche, M., Mamalet, F., Wolf, C., Garcia, C., & Baskurt, A. (2011, November). Sequential deep learning for human action recognition. In *International workshop on human behavior understanding* (pp. 29-39). Springer. 10.1007/978-3-642-25446-8_4

Banjarey, K., Sahu, S. P., & Dewangan, D. K. (2021, April). A Survey on Human Activity Recognition using Sensors and Deep Learning Methods. In *2021 5th International Conference on Computing Methodologies and Communication (ICCMC)* (pp. 1610-1617). IEEE. 10.1109/ICCMC51019.2021.9418255

Bantupalli, K., & Xie, Y. (2018). American sign language recognition using deep learning and computer vision. In *2018 IEEE International Conference on Big Data (Big Data)* (pp. 4896-4899). IEEE. 10.1109/BigData.2018.8622141

Bowers, A. J., & Zhou, X. (2019). Receiver operating characteristic (ROC) area under the curve (AUC): A diagnostic measure for evaluating the accuracy of predictors of education outcomes. *Journal of Education for Students Placed at Risk*, *24*(1), 20–46. doi:10.1080/10824669.2018.1523734

Challa, S. K., Kumar, A., & Semwal, V. B. (2021). A multibranch CNN-BiLSTM model for human activity recognition using wearable sensor data. *The Visual Computer*, 1–15. doi:10.100700371-021-02283-3

Chen, J., Zhang, D., Suzauddola, M., Nanehkaran, Y. A., & Sun, Y. (2021). Identification of plant disease images via a squeeze-and-excitation MobileNet model and twice transfer learning. *IET Image Processing*, *15*(5), 1115–1127. doi:10.1049/ipr2.12090

Dong, C., Leu, M. C., & Yin, Z. (2015). American sign language alphabet recognition using microsoft kinect. In *Proceedings of the IEEE conference on computer vision and pattern recognition workshops* (pp. 44-52). IEEE.

Dua, N., Singh, S. N., & Semwal, V. B. (2021). Multi-input CNN-GRU based human activity recognition using wearable sensors. *Computing*, *103*(7), 1–18. doi:10.100700607-021-00928-8

Garcia, B., & Viesca, S. A. (2016). Real-time American sign language recognition with convolutional neural networks. *Convolutional Neural Networks for Visual Recognition*, *2*, 225–232.

Ghanem, S., Conly, C., & Athitsos, V. (2017, June). A survey on sign language recognition using smartphones. In *Proceedings of the 10th International Conference on PErvasive Technologies Related to Assistive Environments* (pp. 171-176). 10.1145/3056540.3056549

Gui, J., Sun, Z., Wen, Y., Tao, D., & Ye, J. (2020). *A review on generative adversarial networks: Algorithms, theory, and applications.* arXiv preprint arXiv:2001.06937.

Jin, C. M., Omar, Z., & Jaward, M. H. (2016). A mobile application of American sign language translation via image processing algorithms. In *2016 IEEE Region 10 Symposium (TENSYMP)* (pp. 104-109). IEEE. 10.1109/TENCONSpring.2016.7519386

Juba, B., & Le, H. S. (2019). Precision-recall versus accuracy and the role of large data sets. *Proceedings of the AAAI Conference on Artificial Intelligence, 33*(01), 4039–4048. doi:10.1609/aaai.v33i01.33014039

Lee, C. K., Ng, K. K., Chen, C. H., Lau, H. C., Chung, S. Y., & Tsoi, T. (2021). American sign language recognition and training method with recurrent neural network. *Expert Systems with Applications, 167*, 114403. doi:10.1016/j.eswa.2020.114403

McKee, M. M., Paasche-Orlow, M. K., Winters, P. C., Fiscella, K., Zazove, P., Sen, A., & Pearson, T. (2015). Assessing health literacy in deaf American sign language users. *Journal of Health Communication, 20*(sup2), 92-100.

Noroozi, F., Kaminska, D., Corneanu, C., Sapinski, T., Escalera, S., & Anbarjafari, G. (2018). Survey on emotional body gesture recognition. *IEEE Transactions on Affective Computing*.

Powers, D. M. (2020). *Evaluation: from precision, recall and F-measure to ROC, informedness, markedness and correlation.* arXiv preprint arXiv:2010.16061.

Sharma, S., & Tiwari, S. (2021). COVID-19 Diagnosis using X-Ray Images and Deep learning. In *2021 International Conference on Artificial Intelligence and Smart Systems (ICAIS)* (pp. 344-349). IEEE. 10.1109/ICAIS50930.2021.9395851

Srinivasu, P. N., SivaSai, J. G., Ijaz, M. F., Bhoi, A. K., Kim, W., & Kang, J. J. (2021). Classification of skin disease using deep learning neural networks with MobileNet V2 and LSTM. *Sensors (Basel), 21*(8), 2852. doi:10.339021082852 PMID:33919583

Taskiran, M., Killioglu, M., & Kahraman, N. (2018). A real-time system for recognition of American sign language by using deep learning. In *2018 41st International Conference on Telecommunications and Signal Processing (TSP)* (pp. 1-5). IEEE. 10.1109/TSP.2018.8441304 tecperson. (2017). Sign language MNIST. *Kaggle*. Retrieved from https://www.kaggle.com/datamunge/sign-language-mnist

Tiwari, S. (2018a). Blur classification using segmentation based fractal texture analysis. *Indonesian Journal of Electrical Engineering and Informatics*, *6*(4), 373–384.

Tiwari, S. (2018b). An analysis in tissue classification for colorectal cancer histology using convolution neural network and colour models. *International Journal of Information System Modeling and Design*, *9*(4), 1–19. doi:10.4018/IJISMD.2018100101

Tiwari, S., & Jain, A. (2021). Convolutional capsule network for COVID-19 detection using radiography images. *International Journal of Imaging Systems and Technology*, *31*(2), 525–539. doi:10.1002/ima.22566 PMID:33821095

Tiwari, S., Jain, A., Sharma, A. K., & Almustafa, K. M. (2021). Phonocardiogram Signal Based Multi-Class Cardiac Diagnostic Decision Support System. *IEEE Access: Practical Innovations, Open Solutions*, *9*, 110710–110722. doi:10.1109/ACCESS.2021.3103316

KEY TERMS AND DEFINITIONS

ConvNet: CNN or ConvNet is a type of deep neural network that is frequently used to evaluate visual imagery.

Deep Learning: Deep learning is a sort of machine learning and artificial intelligence that mimics how humans acquire knowledge. Data science, which covers statistics and predictive modelling, incorporates deep learning as a key component.

Gesture: A gesture is a visual representation of physical action or emotional expression. It consists of both body and hand gestures.

MobileNet: MobileNet is a refined deep neural network architecture that constructs lightweight CNN model using depthwise separable convolutions and provides an effective model for mobile and embedded application areas.

Sign Language: When spoken communication is not possible, sign language is a method of communicating through body motions, particularly those of the hands and arms.

Transfer Learning: Transfer learning is the concept of breaking free from the isolated learning paradigm and applying what you've learned to solve related problems.

APPENDIX: ABBREVIATIONS

AI Artificial Intelligence
ASL American Sign Language
AUC Area under Curve
CNN Convolutional Neural Network
DBN Deep Belief Network
GAN Generative Adversarial Network
GRU Gated Recurrent Unit
HCI Human Computer Interaction
KNN K-Nearest-Neighbour
LSTM Long-Short Term Memory
MLP Multilayer Perceptron
NN Neural Network
RF Random Forest
RNN Recurrent Neural Network
ROC Receiver Operating Characteristic Curve
SURF Speeded Up Robust Features
SVM Support Vector Machine

Chapter 10
Applications of Hand Gesture Recognition

Hitesh Kumar Sharma
University of Petroleum and Energy Studies, India

Tanupriya Choudhury
 https://orcid.org/0000-0002-9826-2759
School of Computer Science, University of Petroleum and Energy Studies, India

ABSTRACT

Hand gestures, as the name suggests, are different gestures made by the use of hands. Historically, hand gestures have been created to communicate with people who were unable to speak, but as new technologies have emerged, hand gestures have been used widely in different fields such as medicine, defense, IT industry. Hand gestures are being used to create TVs without remotes. Face recognition is also used to verify the user and to change channels, increase/decrease volume. To switch on or off the lights in the house, hand gesture recognition devices are being developed. Hand gesture recognition (HGR) is a natural kind of human-machine interaction that has been used in a variety of settings. In this chapter, the authors have described the application of HGR in various sectors. They have also explained the tools and techniques used for HGR.

INTRODUCTION

Hand gesture recognition is the ability to recognize and interpret movements of human body so as to interact with or control a computer system without any physical touch. Hand gesture recognition is very useful for interaction between

DOI: 10.4018/978-1-7998-9434-6.ch010

humans and any kinds of machine especially computers. Computers don't normally understand hand gestures but can be programmed to do so. Through hand gestures, human beings would be able to instruct the computer without any requirement of physical touch. This would be possible through the use of sensors which would be useful to contemplate the information the user is trying to convey through his/her hand gestures. Just by performing the hand gestures, we can't expect the computer to recognize the meaning of it because computer can't understand it as the only language understandable by computer is binary. So, first we need to make sure that the hand gestures we make are converted into binary language. So we need to create a framework where the computer can recognize the meanings of the various hand gestures after converting into binary and after decoding it, can decide the action to be performed. Hand gesture recognition mainly follow two types of analysis. First is glove based analysis and second is vision based analysis. The glove based method uses the mechanism of sensors (which could be mechanical or optical) which are attached to the glove. These sensors act as a transducer medium which converts change in a physical quantity like pressure, position, brightness or temperature into electrical signals which are easily understood by the computer system. The relative posture of hand is again determined by another sensor which is usually magnetic or acoustic in nature. On the other hand, vision based method uses various types of cameras like monocular fisheye, ToF and IR. These cameras measure the time taken by the light to bounce off of an object and then it uses the delay time to find out the distance of the object. But this method also faces some challenges like lighting variation, background issues, effect of occlusions or processing time traded against frame rate and resolution.

LITERATURE REVIEW

Hand Gesture Recognition (HGR) plays an important role in various sectors mainly in Gaming, Medical and Auto driving and parking system. In their survey (K. Banjarey et al., 2022) presented the various methods to identify various human activity. Based on the movements of hand, the wearable sensor device recognizes hand gestures. The concept is utilized by the Internet of Medical Things (IoMT). With its help it has been easier for the patients as well as the doctors to interact. IoMT has made things easier for the distanced and old patients who cannot visit the hospital on the regular basis. Deep Learning is one of the majorly used approach for hand gesture recognition (Banjarey K et al., 2022). It work as a medium between the patients and the healthcare, by understanding the hand gesture the required service is provided to the distanced and old patients. Along with the medical assistance, IoMT also helps in maintaining the data of number of patients correctly and accurately. Data is stored digitally and this doesn't leads to paper wastage (D. K. Dewangan, 2021). For security purposes, fingerprints or pin locks can be used which also comes under hand gesture recognition. Doctors can detect the cause on their virtual screens. In coming future, if we gain more advancement on hand gesture recognition then it can be possible that surgeries will be performed through gestures. Hand gestures will be processed by the system and after processing the desired output/action will be performed, for example: picking up a tool to start a surgery. It will also enable distancing between the patient and the doctor, as during the pandemic it was risky for the doctors to go near the patient and do his/her check up because of the corona virus. With more advancement, assigning and billing of medicines to the patients can also be done with help of hand gesture recognition.

FEATURE-BASED TECHNIQUES FOR HAND GESTURE RECOGNITION

The Data received is always the raw data or the messy data and to change this raw data into structured data, one has to do Data Cleaning so that ML model can work properly and to do this Feature engineering came into picture which helps in Data Cleaning (T. Mantecón et al., 2016). Feature Engineering is essential and Labour intensive component of ML applications. Most ML performances completely depends on representation of Feature vector.

There has been a great development in the field of hand gestures and hand tracking in the recent studies, people are making use of OpenCV (Kumar Sharma, H et al. 2018) and other libraries and packages for implementing some of the cool and trending problem solving machine learning Algorithms for hand gestures and hand tracking .The most common ongoing research field is on hand gesture recognition of the images using American Sign Language(ASL) images (A. Krizhevsky et al., 2012) .

Methods and Materials

For implementing the hand gesture recognition we used some feature selection, data mining, text classification, feature engineering and using machine learning algorithms and also using the camera sensors and video processing for sensing purposes of our hands .We also use Scale-invariant transform(SIFT), and Histogram of Gradients(HOG). For the classification purposes we have used the machine learning algorithms support vector machine(SVM), K-Nearest Neighbor(KNN), and some Bayesian algorithms (H.-J. Kim et al., 2008).

Data Collection

We have collected the data from the online resources. These data are composed of infrared images by sensors. This dataset contains ten different hand gestures that were gathered from 10 different subjects of which 5 were of male and 5 were of female. (Figure 1)

Figure 1. Hand gesture data sample

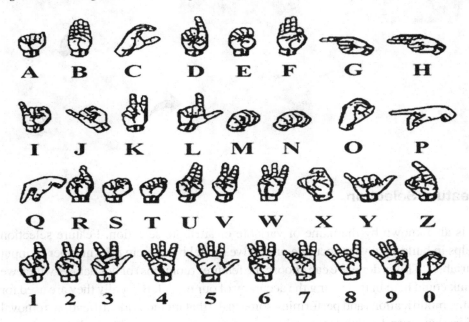

Data Pre-Processing

It is a step or stage in data mining and data analysis which is based on the transformations into a format that can be used for analysis through machine learning algorithms. different forms of data are: Structured tables, Images, Audio files, Videos, etc. As in our datasets we will be having a lot of na values, and 0's and missing values in our dataset that need to be looked at. (Ali A. et al., 2018).

So we can do it like dropping the "na " values, removing the observations(rows) and other fields containing the missing data. So doing or performing such tasks of pre-processing will increases the accuracy and efficiency of our model .In this particular dataset we are using the image datasets in which we have a root folder in which has subfolder of palm, I, fist, moved fist, thumb, index finger, moved palm. (Figure 2)

Figure 2. Hand gesture recognition flow diagram

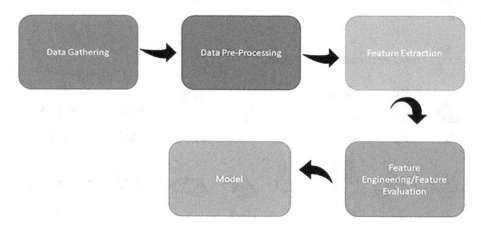

Feature Selection

It is also known by the name of variable or attribute selection. Feature selection helps in building a very accurate predictive model by reducing the number of input variables. This is done We can choose the features (columns) of our dataset which we think could help in increasing the accuracy of our model. Basically they are used for some modifications and performing some tasks that includes identification, removal of irrelevant and redundant attributes or features which might result in giving us the output with less accuracy (C. Szegedy et al., 2015). This becomes an important aspect in order to improve classification performance. Classification Performance can be high if quality data is fed as an input. Some of the examples are correlation coefficient scores, information gain, etc.

Feature Engineering

The process of selecting, transforming and manipulating the raw data into features .Our raw data is being converted into features .We develop new features which is not already existing in our training dataset and will help us in performing fast data transformations which will improve our accuracy of our model .Feature engineering involves various processes: we can add and remove some of the features, plotting and visualizing the data, extracting some of the useful insights and features from our data, Exploratory Data Analysis, running test datasets to check if our model is performing well or not as compared to the benchmark while using boosting and bagging techniques (HK Sharma et al., 2015) . We have some machine learning algorithms: we work with the missing values or data (filtering out na values),we need to handle a lot of outliers that we get after performing visualizations on our data so we need to make sure to remove such outliers as best as possible. Linear regression is helpful for outliers, we need to perform various encoding techniques and scaling as well and some amount of normalizations and standardization (HK Sharma et al., 2015).

Scale Invariant Feature Engineering (SIFT): It involves various methods or techniques:

1.   Image matching
2.   Motion segmentations
3.   Detecting stable or salient features or points in more than two images
4.   Refinement of scale and location
5.   Determining orientations
6.   Determining descriptors

Histogram of Oriented Gradients (HOG): Used mainly in the fields of computer Vision and Image Processing in the case of object detection. Angle of gradient is used for computing the features.

The aim and objective of this supervised learning algorithm is to generate a hyperplane in an N-dimensional space that is used for classifying the data points distinctly. We find out the plane that will be having the maximum margin, the distance between the points being maximum. These are those data points that lie close to the hyperplane. The loss function helps in making the margin maximum (Mais Yasen et al., 2019).

APPLICATIONS OF HAND GESTURE RECOGNITION

Hand Gesture Recognition has multiple uses in the modern day machine infested world. Its uses can be found almost everywhere in our everyday life.The gaming technology is changing and getting better very rapidly, people are getting introduced to virtual/augmented reality and virtual games. In this type of gaming, people can experience enhanced experience as they not really need to physically handle the character controls using a jockey or playstation screen but can play using their hand gestures, which are later decoded by the computer and can be used to control the movements of their character (Nico Zengeler et al., 2019).

Figure 3. Major applications of hand gesture recognition

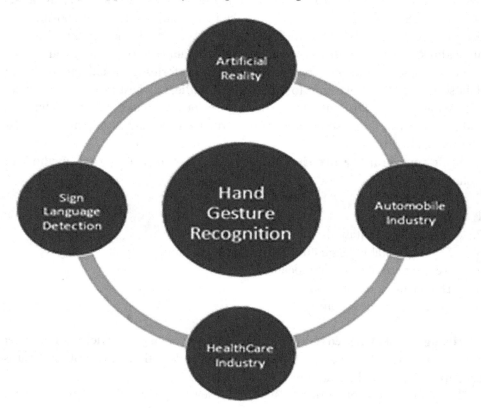

In Sign language recognition: People who are deaf can't understand what is being spoken to them, thus the information needs to be written down or should be represented by pictures. With hand gesture recognition, the computer is able to understand the sign language and is able to interpret the information and could display the given information. Doctors often handle critical cases and thus their time is very precious. With hand gesture recognition devices, the doctor or any medical staff could view the medical reports of the patient without touching it and could conduct various operations like that of the laser surgery for cataract or even complex heart surgeries (Bhushan, A et al., 2017). Our homes are becoming modern and technology adaptable day by day. People often think to buy air conditioners whose flow of air and the temperature or televisions whose channels or volume or washing machines whose washing time and procedure could be controlled by moving one's hand. (Figure 3).

Hand Gesture Movements, something which we always do when we say a good 'bye' to someone. How cool would it be if you just wave your hand towards a camera and it recognizes that you are saying a goodbye to someone? How about you move your palm to the right and the slideshow proceeds to the next slide? All of this seems so exciting and amusing and to your surprise all of this is a reality now. You can make hand gestures and the Algorithms will guess the gestures with a very high accuracy in no time. All of this has been possible by the advancements in technology, especially in the field of Artificial Intelligence and Machine Learning. Now you must be wondering how does this actually work? I've answered this to the best of my ability in the next paragraph. Detecting Hand Gesture Movements comes under the umbrella of detecting Human movements, just like we humans have a network of neurons spread on our entire body, computers have an artificial neuron network which is known as the Convolutional Neural Network (CNN) (S Taneja et al., 2018). The Image is broken into various datasets and the CNN compares the image to be detected with the existing images and outputs the result which is the most common in the new Image and existing dataset. Now you might be wondering that where can we practically use this Gesture Detection Technique? Let's know more about this in next paragraph. (Figure 4)

Figure 4. CNN model based basic diagram for hand gesture recognition

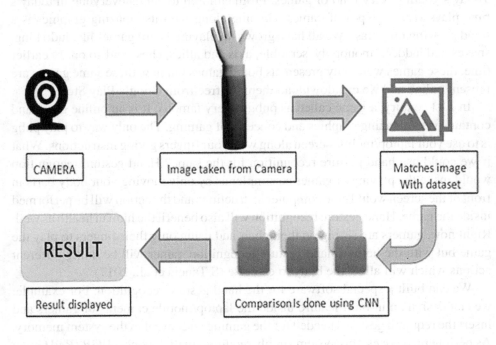

201

Human Robot Interaction

Human Robert Interaction is such a field where hand gesture recognition can play a major role. Even though its implementation has already been introduced in Human Robot Interaction but still some high advancements can be done. In other words, hand gesture recognition is a virtual form of sign language. Through hand gestures we can instruct the robot to perform various tasks. For example: If a human wants a robot to make some coffee, there will be different hand gestures required for the same (Patni, J.C. et al., 2019). The robot will scan the hand gesture, process it into its memory and will give the desired output. The desired output will be given after performing the following steps. First hand gesture will instruct the robot to boil defined quantity of water. Second hand gesture will instruct the robot to add fixed amount of coffee and sugar as prescribed by the human. And with the final hand gesture the robot will serve the coffee. Therefore, with the help of hand gesture recognition it is possible for the robot to perform every possible human day-to-day activities.

Gaming and Graphics

Today's youth is very fond of games. From children to adults everyone in today's time plays various types of games. Gaming along with its amazing graphics is a trend growing up so fast. We all have grown up playing board games like ludo king, snakes and ladders, monopoly, scrabble, axis and allies, chess and so on. In earlier time, these games were only present as board games but now these same games are present online and we can download them for free from Google Play Store.

In 21st century, a game called as pubg is very famous. It is an online game and contains very amazing graphics and concepts of gaming. The only way to play pubg is to use your laptop/mobile screen along with your fingers giving instructions. What if we could use hand gesture recognition for the same? Hand gesture recognition would make the playing of games so simpler as by only moving your body parts in front of the screen would give computer instructions and the action will be performed inside the game. Hand gesture recognition will also benefit the human health as well. Right now, gamers are sitting on their chair and using only their fingers to play the game but with the help of hand gesture recognition gamer will be doing different actions which will allow the body to exercise (S Taneja et al., 2017).

We can built a special software for the hand gesture recognition. For example: we can design an in-built feature inside the laptop/mobile cameras of gamers and insert the required gestures needed for the gaming and save it in the system memory. As peripheral devices, this system simply requires small, low-cost RGB (Red Green Blue) high speed cameras. At the same time, 3D modeling system based on hand gesture recognition also need to be introduced into the system. (Figure 5)

Figure 5. HGR processing in gaming console

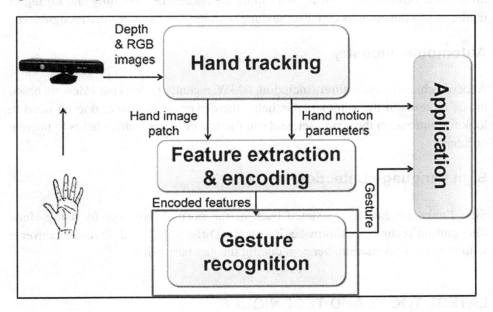

Medical Sector

Hand gesture recognition plays an important role in medical sector. Based on the movements of hand, the wearable sensor device recognizes hand gestures. The concept is utilized by the Internet of Medical Things (IoMT). With its help it has been easier for the patients as well as the doctors to interact. IoMT has made things easier for the distanced and old patients who cannot visit the hospital on the regular basis. It work as a medium between the patients and the healthcare, by understanding the hand gesture the required service is provided to the distanced and old patients. Along with the medical assistance, IoMT also helps in maintaining the data of number of patients correctly and accurately. Data is stored digitally and this doesn't leads to paper wastage. For security purposes, fingerprints or pin locks can be used which also comes under hand gesture recognition. Doctors can detect the cause on their virtual screens. In coming future, if we gain more advancement on hand gesture recognition then it can be possible that surgeries will be performed through gestures. Hand gestures will be processed by the system and after processing the desired output/action will be performed, for example: picking up a tool to start a surgery. It will also enable distancing between the patient and the doctor, as during the pandemic it was risky for the doctors to go near the patient and do his/her check up because of the corona virus. With more advancement, assigning and billing of medicines to the patients can also be done with help of hand gesture recognition.

Automobile Industry

A lot of vehicle manufacturers including BMW, recently added Hand Gesture based music systems in the vehicles, this helps the driver as the driver doesn't need to look for buttons on the stereo set, and can focus on the road which helps to prevent accidents.

Sign Language Detection

Sign Language can be translated back in the normal language by using Hand Recognition Techniques, this makes it easier for Differently Abled citizens to converse with the abled population, hence reducing the gap between us.

LATEST TOOLS AND TECGNIQUES UTILIZED IN HAND RECOGNITION

Gesture recognition is a method derived with the help of computer science and language technology. It helps to translate gestures via bodily movements mostly from face or hands to mathematical formula. It is a method where we don't really need to come in contact with the devices to control them. Our emotions or sign language helps to interact with them by using their cameras. There are different types of hand gesture recognitions, some of them are:

- Vision
- Kinetic
- Real-time
- Static
- Neural network

The furthest goal of this technology is to make interaction between humans and computers as similar as interaction between humans. Due to which gesture recognition, understanding body language and emotions is one of the most important things to make it possible.

Single Camera

A camera is employed for gesture recognition thanks to taking an picture of the gesture made by hand of the user This image is taken as an input of hand .though it's also employed in detecting hand gesture because the way of input . stereo camera employed for the 3D representation of what's is being seen.

Detecting Object

The technique has two requirement . the primary requirement is to detect the image that contains the article or not . the second requirement is to search out the position of the topic of the image . As introduced within the previous section, there are many algorithms that perform the task. During this study, the necessity is that the accuracy of the results further as being fast enough to control for real-time application. Therefore, implementation idea is used to multiscale and window technique to separate the image into ROI.

Tracking Object

When the algorithm has detected the frame containing the item because the ROI area, the subsequent thing is to lock and track the target when it is moving or may be partially deformed within the next frames. The utilization of the tracking object are going to be necessary since the user of gesture will happen in few seconds. If we still use classification and detection techniques to conclude, it will be difficult to attain the required processing speed or it is going to result in false conclusions.

Wi-Fi Sensing

During this, its mouse gesture tracking, where the motion of a mouse is correlated to an emblem being drawn by a persons' hand, which may study the changes over time to represent gesture. The software also compensates for the inadvertent movements.

Wired Gloves

These gloves also provide input to computer of hand gesture by the position and rotation of hands using magnetic or inertial tracking devices some gloves may also detect the bending of fingers. This user optic cables running down the rare of the hand. The primary commercially available hand tracking gloves type device was the info gloves.

CONCLUSION

Hand gesture recognition or HGR is changing our relation with technical devices Classified as touch less user interface (TUI) which is giving rise to a whole new World of input possibilities. With gesture recognition an individual can perform a certain task or an input command Is performed without making any contact with the device. In this a sensor responds to a Movement and then executes a particular command.

This technology has been around for past few years and made our life easier. Automotive manufacturers are already using it in cars to allow the drivers to control Certain things like- music volume, managing calls, switch· on/of lights and etc. To overcome common challenges and achieve a reliable result, it is important to create trustworthy and robust algorithms with the help of a camera sensor that has a certain feature, which needs substantial work. To be sure, each of the techniques listed above is not without its own set of pros and downsides.

REFERENCES

Alani. (2018). Hand Gesture Recognition Using an Adapted Convolutional Neural Network with Data Augmentation. *24th IEEE International Conference on Information Management.*

Banjarey, K., Prakash Sahu, S., & Kumar Dewangan, D. (2021). A Survey on Human Activity Recognition using Sensors and Deep Learning Methods. *2021 5th International Conference on Computing Methodologies and Communication (ICCMC),* 1610-1617. 10.1109/ICCMC51019.2021.9418255

Banjarey, K., Sahu, S. P., & Dewangan, D. K. (2022). Human Activity Recognition Using 1D Convolutional Neural Network. In S. Shakya, V. E. Balas, S. Kamolphiwong, & K. L. Du (Eds.), *Sentimental Analysis and Deep Learning. Advances in Intelligent Systems and Computing* (Vol. 1408). Springer., doi:10.1007/978-981-16-5157-1_54

Bhushan, A., Rastogi, P., Sharma, H. K., & Ahmed, M. E. (2017). I/O and memory management: Two keys for tuning RDBMS. *Proceedings on 2016 2nd International Conference on Next Generation Computing Technologies, NGCT 2016,* 208–214.

Dewangan & Sahu. (2021). Driving Behavior Analysis of Intelligent Vehicle System for Lane Detection Using Vision-Sensor. *IEEE Sensors Journal, 21*(5), 6367-6375. . doi:10.1109/JSEN.2020.3037340

Kim, L., & Park. (2008). Dynamic hand gesture recognition using a CNN model with 3D receptive fields. *International Conference on Neural Networks and Signal Processing*.

Krizhevsky, Sutskever, & Hinton. (2012). *Imagenet classification with deep convolutional neural networks*. NIPS.

Kumar Sharma, H., & Kshitiz, K. (2018). NLP and Machine Learning Techniques for Detecting Insulting Comments on Social Networking Platforms. *ICACCE 2018*.

Mantecón, T., del Blanco, C. R., Jaureguizar, F., & García, N. (2016). Hand Gesture Recognition using Infrared Imagery Provided by Leap Motion Controller. *Int. Conf. on Advanced Concepts for Intelligent Vision Systems, ACIVS 2016*, 47-57. ()10.1007/978-3-319-48680-2_5

Patni, J. C., & Sharma, H. K. (2019). Air Quality Prediction using Artificial Neural Networks. *2019 International Conference on Automation, Computational and Technology Management, ICACTM 2019*, 568–572. 10.1109/ICACTM.2019.8776774

Sharma, Shastri, & Biswas. (2013). A Framework for Automated Database TuningUsing Dynamic SGA Parameters and Basic Operating System Utilities. *Database System Journal*.

Sharma, Shastri, & Biswas. (2015). Auto-selection and management of dynamic SGA parameters in RDBMS. *Database System Journal*.

Szegedy, L., Jia, S., Reed, A., & Erhan, V., & Rabinovich. (2015). Going deeper with convolutions. *2015 IEEE Conference on Computer Vision and Pattern Recognition (CVPR)*.

Taneja, S., Karthik, M., Shukla, M., & Sharma, H. K. (2017). *AirBits: A Web Application Development using Microsoft Azure*. International Conference on Recent Developments in Science, Technology, Humanities and Management (ICRDSTHM-17), Kuala Lumpur, Malaysia.

Taneja, Karthik, Shukla, & Sharma. (n.d.). Architecture of IOT based Real Time Tracking System. *International Journal of Innovations & Advancement in Computer Science, 6*(12).

Yasen, M., & Jusoh, S. (2019, September). A systematic review on hand gesture recognition techniques, challenges and applications. *PeerJ. Computer Science, 5*, e218. doi:10.7717/peerj-cs.218 PMID:33816871

Zengeler, N., Kopinski, T., & Handmann, U. (2019). Hand Gesture Recognition in Automotive Human–Machine Interaction Using Depth Cameras. *Sensors (Basel), 19*(1), 59. doi:10.339019010059 PMID:30586882

Chapter 11
Application of HGR Rock– Paper–Scissors Model

Keshav Garg
School of Computer Science, University of Petroleum and Energy Studies, India

Prabhjot Singh
School of Computer Science, University of Petroleum and Energy Studies, India

Bhupesh Kumar Dewangan
iD https://orcid.org/0000-0001-8116-7563
School of Engineering, Department of Computer Science and Engineering, O.P. Jindal University, Raigarh, India

ABSTRACT

Although great progress has been made by leveraging the use of sensors in various fields like human body tracking, robust, accurate, and efficient hand gesture recognition still remains an open problem. The authors draw a comparison on how the HSV (hue, saturation, value) model used in hand gesture recognition is better than the other models. Also, the focus will be laid on the different segmentation techniques that are being used nowadays and how thresholding is an important part of it. The extraction of noisy hand contours is really challenging, and for the same reason, the authors propose the usage of convexity defects to record deviation from contours and the dilution and erosion of the binary images to recognize hand gestures like rock-paper-scissors. This model has been tried and tested on different devices with various hand gestures and yields a positive result.

DOI: 10.4018/978-1-7998-9434-6.ch011

INTRODUCTION

The term Human-Computer Interaction(HCI) came into being in the early 1980s. It was made famous by Stuart K. Card, Allen Newell, and Thomas P. Moran in their seminal 1983 book, "The Psychology of Human-Computer Interaction". With the development of new technologies and the studies of various researchers, Human-Computer Interaction has reached great heights. It has been in trend for the past few years and has been growing rapidly since then.

The continuous development in this area gave rise to various fields in this sector. The prominent ones include:

- **Gesture Recognition** - An ingenious way of controlling various functions of a device like a computer using human gestures is known as gesture recognition. Example: Controlling the volume of the speakers of a computer using Hand Gestures as shown in Fig.1.

Figure 1. Adjusting volume using hand gestures

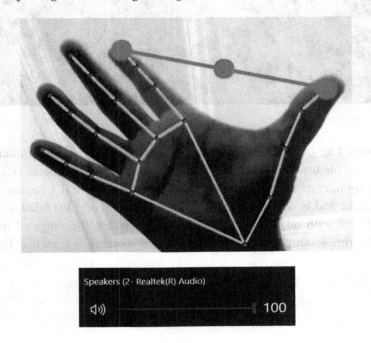

- **Object Tracking** - Object Tracking involves tracking the motion state of a particular object, such as a person, to read their anatomy, detect their location, record and analyze their activities, etc., in each frame of an image sequence or a video. Example: Monitoring the traffic of a particular area, i.e. the number of vehicles, their types, and their speed (Lee, Y. H.,et al.,2019) as shown in Fig.2.

Figure 2. Traffic monitoring system

- **Natural Language User Interface**(LUI or NLUI): NLUI is an interface that provides the user with the liberty to interact with the computer system using human-understandable language. Unfortunately, Natural Language is quite uncertain and is dependant on a vast amount of world knowledge. It can be combined with other ways of interaction to widen the range of interaction bandwidth. Example: Dialog System whose working is shown in Fig.3.

Figure 3. Architecture overview of dialog management system
(Wachtel, A, et al., 2017)

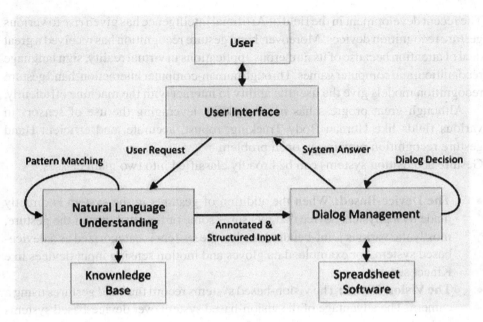

The future of HCI is vast and promising since we have seen the world go from keyboard to mouse point-and-click, to touch, and now we're into the world of gesture and voice recognition, which in itself is a remarkable evolution for Human-Computer Interaction. As far as HCI is concerned, hand gesture recognition plays a significant role in it and has a variety of applications. The development in the field of hand gesture recognition systems also avoided diseases such as Parkinson's disease, trigger finger, gorilla arm, etc., which is caused by excessive use of the mouse or keyboard (Rautaray, S. S., & Agrawal, A. (2015)). The main aim of developing a hand gesture recognition system is to establish an interaction between humans and computers from the recognized gestures. Many vision-based and non-vision-based methods have been proposed in recognizing hand gestures and extracting information from them, as discussed under the section of comparative analysis.

This chapter focuses on the different approaches in hand-gesture recognition; segmentation in the HSV model; comparison on how the HSV(Hue, Saturation, Value) model is better than the other models; challenges faced during gesture recognition; and The ROCK-PAPER-SCISSORS GAME Model explanation.

HAND GESTURE RECOGNITION: RELATED WORK

The recent development in the field of Artificial intelligence has given rise to various gesture recognition devices. Moreover, hand gesture recognition has received a great deal of attention because of its numerous applications in virtual reality, sign language recognition, and computer games. Through human-computer interaction, hand gesture recognition models give the user the ability to interact with the machine efficiently.

Although great progress has been made by leveraging the use of sensors in various fields like Human Body Tracking, robust, accurate, and efficient Hand gesture recognition remains an open problem.

Gesture acquisition systems can be broadly classified into two main classes,

- **The Device-Based:** When the addition of gestures in the system is directly made by a physical device that measures some of the attributes of the gesture, mostly the various joint-bending angles, the system is categorized as a device-based system, for example, data gloves and motion sensing input devices like Kinect sensors.

- **The Vision-Based:** The vision-based systems record the hand gestures using a camera. The advantage of the vision-based system over device-based systems is due to its unconstrained nature. Its main drawback is due to its complexity of processing, whereas the device-based capturing is fast and robust.

In developing a gesture recognition system, the gestures can be static, which requires less computation, or dynamic, which can be described as a sequence of postures that are more complex and require complex computation but are suitable for a real-time environment. In the case of a dynamic gesture, by treating it as the output of a stochastic process, hand gesture recognition can be addressed based on statistical modellings, such as PCA, HMMs, and more advanced particle filtering and condensation algorithms. Dynamic gestures can also be recognized using the technique of depth matrix and adaptive-based classifier (Kane, L., & Khanna, P., 2019). In this work, the researcher recognized the postures using depth matrix and the 1-nearest neighbour strategy. Posture sequence labels are predicted by a dynamic naive Bayes classifier which works in association with an adaptive windowing mechanism. This method was a fast alternative to region-based descriptors.

Several other methods have been observed in capturing the input from the user. The commonly used ones include data gloves, hand belts, and cameras. Unlike optical sensors, data gloves are more reliable in gesture recognition systems (Kim, J. H.,et al.,2009) (Haria, A.,et al,2017). But this method requires the user to wear data gloves every time, which makes it inconvenient. Also, data gloves are more expensive than optical sensors, for example, cameras. For this reason, the recent development of inexpensive depth cameras, for example, Kinect sensors, brought new opportunities for hand gesture recognition.

But this does not completely solve the background noise problem to handle noisy hand gesture shapes obtained from Kinect sensor (Ren, Z.,et al., 2017). In this work, the researcher proposed a method to measure similarity between hand shapes using Finger-Earth Mover's Distance metric. Most of the gesture recognition systems extract the ROI (Region of interest) using colour spaces. The inbuilt cameras of the devices do not provide depth information like depth cameras, but they work with less computing costs. Therefore, to overcome the high expenditure, the authors in their model used a built-in webcam of the laptop, unlike the use of any additional cameras or hand markers such as gloves.

After the gesture is recorded, the preprocessing of the images is conducted using various algorithms and strategies such as noise removal, edge detection, smoothening, and sharpening, which is then followed by different segmentation techniques for boundary extraction, i.e. separating the foreground from the background. Using a Gaussian filter, the smoothing of the extracted image is done with the appropriate size of the kernel passed. To eliminate the noise in the image, morphological transformation techniques of erosion and dilution are used, which is usually performed on binary images.

Some of the organizations using hand gesture recognition models are mentioned below:

- **GestureTek™:** The organization's multi-patented video gesture control innovation (VGC) allows clients to control multi-media content, access data, control special effects, even drench themselves in an intuitive 3D virtual world – basically by moving their hands and body.
- **Microsoft and Intel:** In 2019, tech giants Microsoft and Intel started developing use cases for gesture recognition. Intel delivered a white paper on touchless multifaceted authentication (MFA) for use by healthcare organizations to alleviate security risks and further develop clinician effectiveness. On the other hand, Microsoft has a project to investigate camera-based gesture recognition within surgical settings. Also, Microsoft's Xbox360 captures body and hand motions in real-time, freeing gamers from keyboards and joysticks.
- **Google:** Even Google has launched hand-gesture recognition features in its smartphones that don't rely on any expensive depth sensors or processor-intensive high-priced technology. This technology is also scalable to several hands simultaneously.
- **Bavarian Motor Works (BMW):** Since 2016, the 7 Series of vehicles have had gesture recognition features that permit drivers to turn up or down the

volume, accept or reject a call, and change the point of view of the multi-camera. The additional customizable two-finger gesture is also accessible.

- **Sign Language and Cognition:** Along with these organizations, the hand gestures for communicating in various languages for the speakers as well as the listeners and the gestures for cognition are being used worldwide (Clough, S., & Duff, M. C.,2020). It's been quite a while that sign language has become a part of human civilization, even before the verbal discussions. Now the sign language has taken a special place in the fields of especially disabled, armed forces, and air traffic controls. It is also extremely helpful in the road and long-distance communications. Gestures can be termed as part of initial forms of communication to express feelings and needs with expressions. The analysis of gestures is an ideal way of connecting with thoughts and actions, and gestures influence cognition because they facilitate this action. Hence, gestures can predict children's readiness as well as help in the development of the child in various domains. By better grasping the manner in which gestures impact thinking, particularly across development, we could all the more likely comprehend the association between activity, insight, and thinking, and as a result, the manner in which cognition is grounded. Thus, Sign Language and Cognition are also important uses of Gesture Recognition Systems.

A model has been proposed by researchers in which the challenging task of recognition of fingerspelling postures due to complex computation was made time efficient using modified shape matrix variants on depth silhouettes. In this work, the hand postures were acquired using depth sensors (Kane, L., & Khanna, P.,2015).

- **Other Smart Home Appliances:** Gesture Recognition is used in various other smart devices used in day-to-day activities like in computers, televisions, gaming consoles, fans, radio, etc. In these devices, dynamic gestures are recorded and analyzed to implement various functions corresponding to the device, such as changing the temperature, volume, channels, and so on. A model has been proposed for Hand Gesture Recognition and Interface via a Depth Imaging Sensor for Smart Home Appliances that helps control appliances in smart home environments. (Dinh, D. L., et al., 2014)
- **Smartwatches** have been in trend a lot lately, and major smartphone manufacturers such as Apple and Samsung, along with several other companies, have released their smartwatches recently. Since the watch is worn on the wrist and provides a rich user interface to interact via touch or voice, it can be used to analyze the movements of the user's arm, hand, and even fingers using a variety of sensors. Nowadays, most of these watches have built-in sensors like accelerometers and gyroscope sensors that can be

used to identify the user's gestures with ease. The main challenge is to detect the finger's gestures, and for the same reason, researchers have carried out experiments and tests that have led to positive outcomes in this case study. (Xu, C., et al, 2015)

Recently the development of a new device has gained the attention of several researchers that use wearable biosensors and artificial intelligence (AI) to recognize hand gestures from electrical signals in the forearm. This device can be likely used to control prosthetic limbs using hand gestures. The algorithm has supposedly been taught to recognize 21 different hand gestures, some of them being fist, palm, etc. All this is done using an electronic chip which reduces the computing time and also maintains the user's privacy. If such devices are, in the future, used and implemented in the commercial sector, it will bring a great change to this hi-tech world.

THE CONCEPT OF COLOR IMAGE SEGMENTATION USING HSV

HSV corresponds to Hue, Saturation, and Value. It is also often termed as HSB, i.e. Hue, Saturation, and Brightness. The three fundamental concepts of HSV are:

Hue: Hue is nothing but the colour that shall be denoted as a point in a 360-degree colour circle, as shown in Fig.4 below. In simple words, Hue is the dominant colour as perceived by an observer. Based on the Hue values, it is divided into six parts of 60 degrees from 0 to 360, and their respective colours are mentioned in Table.1.

Figure 4. HSV color space wheel

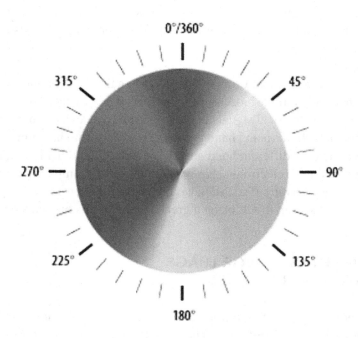

Table 1. Hue values and their respective colors

Red -> 0-60	Cyan -> 181-240
Yellow -> 61-120	Blue -> 241-300
Green -> 121-180	Magenta -> 301-360

Saturation: The amount of white light mixed with Hue is known as Saturation. It is directly linked to the intensity of the colour (range of grey in the colour space). It is normally represented in terms of percentage. The range is from 0 to 100%, where 100% signifies intense colour presence.

Value: Value is the chromatin notion of intensity. It is also termed Brightness, and just like Saturation, it is represented as a percentage. The range is from 0 and goes up to 100%, where 0 denotes black and 100 represents the brightest.

Segmentation is the process of partitioning a digital image into multiple regions and extracting the meaningful region, which is known as Region of Interest (ROI). The region varies with applications, and as a matter of fact, no single universal segmentation algorithm is perfect or even exists for segmenting ROI in all images. Therefore, various segmentation algorithms need to be applied, and the algorithm that gives us the best result is identified.

Image Segmentation is the process of dividing an image into its constituent regions or objects. It is based on two principles:

Similarity Principle - The objective of this principle is to group the pixels based on common property to extract a systematic region. This is also known as Region Approach.

Discontinuity Principle - The objective of this principle is to extract regions that distinguish in properties like colour, intensity, etc. This is also known as Boundary Approach.

The image segmentation algorithms can be classified,

- Based on User Interaction
  - Manual
  - Automatic
  - Semi-Automatic
- Based on Pixel Relationship
  - Contextural (region-based or global)
  - Non-contextual (pixel-based or local)

Image segmentation has a wide range of uses in the field of medicine, object detection, traffic control systems, etc. To be useful, the technique must be combined with a domain's specific knowledge to effectively solve the domain's segmentation problems.

Colour images provide far more information than grayscale images, and for the same reason, the segmentation of coloured images is more reliable. This is mainly done using one of the three colour spaces like RGB, HSV, and YCbCr. The HSV colour space is used mainly by users who wish to select colours from a colour wheel since it correlates better to the experience of colour than the RGB colour space does. As the values of hue lie in the range 0 to 1.0, the respective colour varies from red to yellow, green, cyan, blue, magenta, and then back to red, concluding that both the values 0 and 1.0 indicate red. Similarly, saturation varies from 0 to 1.0, and the corresponding colours vary from unsaturated to fully saturated. The brightness varies from 0 to 1.0, and the brightness of the corresponding colours grows. The first step to segment the coloured image using HSV colour space is to select the Region of Interest (ROI) from the given image, followed by adjusting the HSV values in the ROI to extract the foreground image. Thereafter, the three components H, S, and V are separated and are adjusted to the optimum values. The background is masked to segment the selected object of interest after which, the adjustments of the values of

H, S, and V can be stopped, and the extraction of the foreground image is carried out. This image can be then used for the additional processes of feature extraction and other classification techniques. (Hema, D., & Kannan, D. S., 2020)

COMPARATIVE ANALYSIS

When we look at the human body, the hand is just a small object with several complex articulations and, thus, is more easily affected by segmentation errors. This makes it harder and more challenging to recognize hand gestures.

Segmentation is the most initial process in gesture recognition for recognizing hand gestures. Firstly the hand should be located. For this purpose, two main approaches are used;

First, dividing the video into different frames and each frame to be processed alone. The hand frame is treated as a posture and segmented. The other is using some tracking information such as shape, skin colour. The widely preferred suggestion for hand segmentation is the skin colour since it is easy and unvaried to scale, translation, and rotation changes. The different strategies and tools that are being used during segmentation prefer skin and non-skin pixels to model the hand. But this segmentation technique is most affected by the changes in illumination conditions. To overcome this problem, some researchers presented the use of data gloves and coloured markers which provides additional information about the orientation and position of palms and fingers. But these methods, because of expensive cost and inconvenience, are less preferred in simple applications.

With the upcoming developments of depth sensors such as Kinect sensors, a robust solution to hand segmentation has been provided. The success of the segmentation was further assured by using colour spaces. However, colour spaces are sensitive to lighting changes. Some factors hinder the segmentation process: backgrounds with irregularities, lighting changes, and low video quality.

The author used the colour space technique of HSV to segment the skin colour from the frame and then applied operations like edge detection and normalization to preprocess it.

Now to extract features from recognized hand gestures in different conditions has become challenging. For the same, many methods have been proposed to extract useful information from the images to recognize a particular gesture which have been mentioned in Table 3.

Table 2. Comparative analysis of gesture recognition methods

Blob Detection	ANN (Artificial Neural Network)	Multiclass SVM	Finger Counting Method
It is the process in which the centre of the image captured is selected, and the nearby points are detected that differ in the properties. In one research(Ramjan, M. R.,et al.,2014), this simple method for hand gesture recognition to extract features and templates were used to extract features and Template/Pattern Matching to recognize dynamic hand. The researchers used a low-cost tool to implement the method to type on notepad using hand gestures in different languages. This method of gesture recognition has certain limitations like background and lighting conditions. Another research was conducted (Jalab, H. A., 2012) using this method concluded that efficiency can be improved by background subtraction, so the system can detect hands easily.	In one of the research (Basak, S., & Chowdhury, A., 2014), the static hand gesture was recognized using a neural network. The proposed method used was Wavelet network and ANN(Artificial neural network) for feature extraction.	Another research(*Nagarajan, S., & Subashini, T. S.,2013*) made which classified gestures using the machine learning algorithm SVM.	The Finger counting method is also used to extract certain features from computer vision techniques, specifically contour, convex-hull, and convexity defects for static hand gesture recognition.

Continued on following page

Table 2. Continued

Blob Detection	ANN (Artificial Neural Network)	Multiclass SVM	Finger Counting Method
It is the process in which the centre of the image captured is selected, and the nearby points are detected that differ in properties. In one research(Ramjan, M. R.,et al.,2014), this simple method for hand gesture recognition to extract features and templates were used to extract features and Template/Pattern Matching to recognize dynamic hand. The researchers used a low-cost tool. to implement the method to type on notepad using hand gestures in different languages. This method of gesture recognition has certain limitations like background and lighting conditions. Another research was conducted using this method concluded that efficiency can be improved by background subtraction, so the system can detect hands easily.	In the testing step, 60 images of a hand gesture with various Gaussian noises are used and divided into six classes. The experiment (Basak, S., & Chowdhury, A., 2014), concluded that the overall accuracy boosted up to 97%.	SVM is a linear classifier based on the maximization of the margin between two sets of data. SVM was developed for binary classification, which was extended for multi-class. In multi-class SVM, multiclass labels are decomposed into several two-class labels. It trains the classifiers to solve the problems, and the solution of multi-class problems is reconstructed from the outputs of classifiers.	If fingertips are considered convex points, the trough between the fingers can be regarded as convexity defects. The finger counting method is able to detect the number of fingers effectively and efficiently.

Continued on following page

Table 2. Continued

Blob Detection	ANN (Artificial Neural Network)	Multiclass SVM	Finger Counting Method
It is the process in which the centre of the image captured is selected, and the nearby points are detected that differ in the properties. In one research(Ramjan, M. R.,et al.,2014), this simple method for hand gesture recognition to extract features and templates were used to extract features and Template/Pattern Matching to recognize dynamic hand. The researchers used a low-cost tool to implement the method to type on notepad using hand gestures in different languages. This method of gesture recognition has certain limitations like background and lighting conditions. Another research was conducted (Jalab, H. A., 2012), using this method concluded that efficiency can be improved by background subtraction, so the system can detect hands easily.	To train every input image for classification using machine learning requires so much time and computing resources. The more testing data used, the higher overall accuracy could be generated.	In the research *(Nagarajan, S., & Subashini, T. S.,2013), it was* concluded that Multiclass SVM could provide high accuracy in classifying the hand gestures up to 93.75% *(Nagarajan, S., & Subashini, T. S.,2013).*	This method is fast and easy to implement for hand gesture recognition.

It is important to note that these technologies have different advantages and disadvantages and can be used simultaneously to suit users' requirements. Thus it was concluded that computer vision techniques are swift and easy to implement for hand gesture recognition but still neither flawless nor properly applicable in different situations.

ROCK-PAPER-SCISSORS MODEL

A majority of people grew up playing the traditional game of The Rock-Paper-Scissors with their family and friends. Imagine a model where the users just had to show the gesture and then sit back and relax while the computer itself provides them with the winner of the rounds. They could play as many rounds as they would wish to, without worrying about forgetting the score and even getting notified just in case the opponent had played an unfair match.

The proposed model aims to design a Rock, Paper, and Scissors Game. The two players will show their hand gestures on the screen, and the computer will detect the gesture using computer vision techniques and declare the winner of the round. The rule is rock breaks scissors; scissors cut paper, and paper wraps rock.

Algorithm

```
CALL cv2.imwrite(img1, firstp) to store Player 1 gesture in
.png format in the system
CALL cv2.imread(Location in the system) to read img1 from the
system
CALL cv2.GaussianBlur(firstp) to remove background noise from
the image
SHOW recorded image by calling cv2.imshow(firstp)

CALL cv2.cvtColor(firstp,cv2.COLOR_BGR2HSV) to determine hsv
image for first player's gesture
OBTAIN lower_skin and upper_skin values for the hsv image
CALL cv2.inRange(hsv1,lower_skin,upper_skin)to obtain binary
image for first player's gesture
OBTAIN diluted and eroded image to further remove the
background noise from the image
CALL cv2.threshold(erosion1,cv2.THRESH_BINARY) to obtain
threshold image
SHOW threshold image by calling cv2.imshow('Firstp
threshold',the1)
DETERMINE contours from the image by calling cv2.
findContours(the1.copy(),cv2.RETR_TREE,cv2.CHAIN_APPROX_SIMPLE)
SET max_area to -1
FOR i in range of contours1
  INIT area to cv2.contourArea(contours1[i])
```

```
 IF area > maxarea THEN
    INIT cnt to contours1[i]
    SET max_area to area
SET cnt to cv2.approxPolyDP(cnt,0.01*cv2.
arcLength(cnt,True),True)
CALL cv2.drawContours(firstp, cnt) to draw the contours
SHOW contoured image by calling cv2.imshow("Contoured_
image1",firstp)
OBTAIN convex hull by calling cv2.convexHull(cnt,returnPoints=
False)
OBTAIN convex defects by calling cv2.convexityDefects(cnt,hull)
INIT count_defects to 0
FOR i in range of defects.shape[0]
 CALCULATE angle of hand using mathematical functions
 IF angle <= 90 THEN
 SET count_defects = count_defects+1
 CALL cv2.circle() and cv2.line()
INIT p1_gesture to ""
IF count_defects = 0 THEN
 INIT p1_gesture to "Rock"
 CALL cv2.putText(frame, "Rock") to display Rock on the
mainframe
ELSE IF count_defects = 1 THEN
 INIT p1_gesture to "Scissor"
 CALL cv2.putText(frame, "Scissor") to display Scissor on the
mainframe
ELSE IF count_defects = 4 THEN
 INIT p1_gesture to "Paper"
 CALL cv2.putText(frame, "Paper") to display Paper on the
mainframe
ELSE
 Do nothing
REPEAT
 The same for Player 2 gesture
UNTIL the gesture is recognized using convexity defects
IF p1_gesture = "Rock" and p2_gesture = "Scissor" or p1_gesture
= "Scissor" and p2_gesture = "Paper" or p1_gesture = "Paper"
and p2_gesture = "Rock" THEN
 DISPLAY "Player 1 wins!"
ELSE IF p2_gesture = "Rock" and p1_gesture = "Scissor" or p2_
```

```
gesture =  "Scissor" and p1_gesture = "Paper" or p2_gesture =
"Paper" and p1_gesture = "Rock" THEN
 DISPLAY "Player 2 wins!"

ELSE IF p1_gesture = p2_gesture THEN
 DISPLAY "It's a Draw!"
ELSE
 DISPLAY "Error recording the gesture, try again!"
```

The flowchart for easy understanding of this algorithm is given in Figure 5.

Figure 5. Working of the model

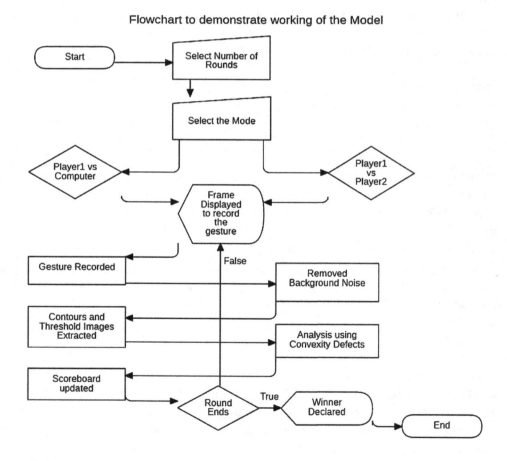

Flowchart to demonstrate working of the Model

Removing Background Noise: Firstly, the background noise for the recorded images is removed using the Gaussian Blur feature. The Gaussian filter is a low-pass filter that removes the high-frequency components, preserves the low-spatial frequency, and reduces the noise, thereby smoothing the image.

```
firstp=cv2.GaussianBlur(firstp,(5,5),0)
secondp=cv2.GaussianBlur(secondp,(5,5),0)
```

firstp and secondp are the gestures recorded for the first player and second player respectively

This is followed by the removal of noise from the binary images of the recorded gestures by using the dilation and erosion functions. Erosion and Dilation are morphological operations used to locate the intensity bumps in an image, isolating the individual elements and combining the distant ones and thus helping to remove noise.

```
kernel=np.ones((3,3),np.uint8)
dilation1=cv2.dilate(first_binary,kernel,iterations=1)
erosion1=cv2.erode(dilation1,kernel,iterations=1)
```

kernel filter helps in smoothing and sharpening of the image by removing the pixels that are darker than a certain fragment of the darkest adjacent pixel

Designing Contours and Threshold Images: Before extracting the contours, thresholding is done to convert the grayscale or coloured image to a binary image for a more straightforward analysis. Then the contours are drawn since the boundary points in the image are known. The green line in Fig.6 depicts the drawn contours.

Figure 6. Contours for scissor, rock, and paper

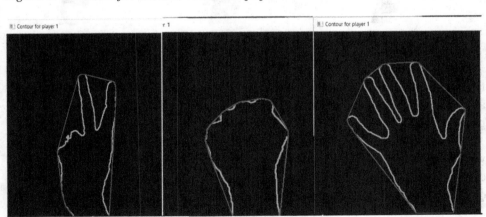

Analysis using Convex Hull and Convexity defects: In Figure 6, the red line denotes the convex hull, i.e. the boundary of the recorded gestures. Any deviation of the contour from the convex hull results in convexity defects, which are then examined and the gesture is displayed on the main frame as Rock, Paper, or Scissors.

```
hull=cv2.convexHull(cnt,returnPoints=False)
defects=cv2.convexityDefects(cnt,hull)
```

Pass returnPoints = False while finding the convex hull in order to find convexity defects

RESULT AND ANALYSIS

In the author's gesture recognizing model, they have included 3 static gestures that are shown in Fig.7.

Figure 7. Gestures for scissor, paper, and rock

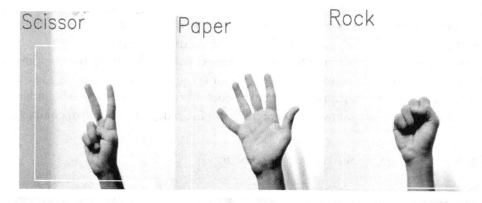

The first gesture on the left, portraying a fist with the index finger and the middle finger extended, forming a "V", denotes the gesture "Scissor". The middle one is an open palm and denotes "Paper". The last one with the closed fist represents the gesture "Rock".

During the implementation of the model, the major drawback and challenge were to minimize the background noise that resulted in the extraction of distorted contours and several other inconsistencies. Due to the same reason, the authors decided to utilize thresholding, convex hull, and convexity defects to detect the hand gestures. The combination of these strategies enabled them to accomplish a better, precise, and efficient model with minimized inconsistencies. To predict the efficiency and accuracy of their model, they conducted two tests. In the primary test, they used conditions with minimal background noise, whereas, in the subsequent test, a background with several irregularities was used.

The average of the number of times a particular gesture was recognized correctly was taken as the accuracy percentage, and the results of the same are mentioned in Table.2.

Table 3. Accuracy of each gesture with a plain background and cluttered background

Gesture	Robustness With a Plain Background (in %)	Robustness With Cluttered Background (in %)
Rock (Convexity Defects=0)	95	93
Paper (Convexity Defects=4)	95	65
Scissor (Convexity Defects=1)	91	60

When executed against any plain background, the model was robust, performed with great efficiency, and without any irregularities. In situations where the background was not plain, the contoured image and convex hull were affected, leading to an unwanted output and thus reducing the model's efficiency. Thus, after being tried and tested, the model is recommended to be used in conditions with minimum background noise to yield the best and positive outcome.

In the future, the accuracy of this model will be improved further, and various other gestures will be added in the same to implement more functions. Moreover, the model currently only tests static gesture images and can be further used in recognizing dynamic gestures in real-time. This simplified model of the hand gesture recognition system will, in time, be used in a variety of hardware such as mobile devices and televisions and also to help the handicapped.

CONCLUSION

The chapter successfully presented the idea of the use of hand gestures as the user interface and the various challenges associated with gesture recognition. Though hand gesture recognition is just a small division of Human-computer interaction, it is one of the most commonly used technologies in this area. While working on the HGR Rock-Paper-Scissors Model, the authors observed how fundamental computer vision techniques can be used for segmentation. Based on the number of convexity defects, the gesture can be classified according to the model. Following the same techniques, the authors were able to build up a robust and efficient gesture recognition system without utilizing the high-cost sensors, i.e. via web camera, hence making it more user-friendly. The overall accuracy of the proposed method reached up to 93.66%. The accuracy is affected by lighting conditions that hinder the detection of skin colour. The lower the background noise and the more refined the illumination conditions, the higher is the accuracy of the hand gesture recognition system. The proposed method proves to recognize hand gestures easily. However, the precision depends on how accurate the object detection phase is implemented. The study of Hand gesture recognition systems is moving at a tremendous speed for futuristic products and services. It is on its way to bringing up a revolutionary change in everyone's life. Its applications include Robotics, Games, Directional Indication through Pointing, and other desktops and smartphone applications. Vast tech giants like Microsoft, Samsung & Sony have already started implementing this technology in their various devices. It is clear that this technology will pave its way into the industries of Education, Entertainment, Machine Learning and Artificial Intelligence, and Medicine in the future. With further research and development in this area, Gesture recognition will emerge as the most cost-effective and efficient way of handling products and services. Gesture Recognition is already in the picture when we see Smart TVs and smartphones being controlled by gestures for implementing various features. In the field of medicine, Hand Gestures may likewise be used in robotic nurses and other medical assistance. As technology is rapidly changing, what is to come is hardly predictable. Still, the fate of Gesture Recognition is undoubtedly digging in to play an increasingly important role in human-computer interaction in the future.

REFERENCES

Basak, S., & Chowdhury, A. (2014). A vision interface framework for intuitive gesture recognition using color based blob detection. *International Journal of Computers and Applications*, *90*(15).

Clough, S., & Duff, M. C. (2020). The role of gesture in communication and cognition: Implications for understanding and treating neurogenic communication disorders. *Frontiers in Human Neuroscience*, *14*, 323. doi:10.3389/fnhum.2020.00323 PMID:32903691

Dinh, D. L., Kim, J. T., & Kim, T. S. (2014). Hand gesture recognition and interface via a depth imaging sensor for smart home appliances. *Energy Procedia*, *62*, 576–582. doi:10.1016/j.egypro.2014.12.419

Haria, A., Subramanian, A., Asokkumar, N., Poddar, S., & Nayak, J. S. (2017). Hand gesture recognition for human computer interaction. *Procedia Computer Science, 115*, 367–374. doi:10.1016/j.procs.2017.09.092

Hema, D., & Kannan, D. S. (2020). *Interactive color image segmentation using HSV color space*. Science and Technology Journal.

Jalab, H. A. (2012). Static hand Gesture recognition for human computer interaction. *Information Technology Journal, 11*(9), 1265–1271. doi:10.3923/itj.2012.1265.1271

Kane, L., & Khanna, P. (2015). A framework for live and cross platform fingerspelling recognition using modified shape matrix variants on depth silhouettes. *Computer Vision and Image Understanding, 141*, 138–151. doi:10.1016/j.cviu.2015.08.001

Kane, L., & Khanna, P. (2019). Depth matrix and adaptive Bayes classifier based dynamic hand gesture recognition. *Pattern Recognition Letters, 120*, 24–30. doi:10.1016/j.patrec.2019.01.003

Kim, J. H., Thang, N. D., & Kim, T. S. (2009, July). 3-D hand motion tracking and gesture recognition using a data glove. In *2009 IEEE International Symposium on Industrial Electronics* (pp. 1013-1018). IEEE. 10.1109/ISIE.2009.5221998

Lee, Y. H., Ahn, H., Ahn, H. B., & Lee, S. Y. (2019). Visual object detection and tracking using analytical learning approach of validity level. *Intelligent Automation and Soft Computing, 25*(1), 205–215.

Nagarajan, S., & Subashini, T. S. (2013). Static hand gesture recognition for sign language alphabets using edge oriented histogram and multi class SVM. *International Journal of Computers and Applications, 82*(4).

Rahmat, R. F., Chairunnisa, T. E. N. G. K. U., Gunawan, D. A. N. I., Pasha, M. F., & Budiarto, R. A. H. M. A. T. (2019). Hand gestures recognition with improved skin color segmentation in human-computer interaction applications. *Journal of Theoretical and Applied Information Technology, 97*(3), 727–739.

Ramjan, M. R., Sandip, R. M., Uttam, P. S., & Srimant, W. S. (2014). Dynamic hand gesture recognition and detection for real time using human computer interaction. *International Journal (Toronto, Ont.), 2*(3).

Rautaray, S. S., & Agrawal, A. (2015). Vision based hand gesture recognition for human computer interaction: A survey. *Artificial Intelligence Review, 43*(1), 1–54. doi:10.100710462-012-9356-9

Ren, Z., Yuan, J., Meng, J., & Zhang, Z. (2013). Robust part-based hand gesture recognition using kinect sensor. *IEEE Transactions on Multimedia, 15*(5), 1110–1120. doi:10.1109/TMM.2013.2246148

Wachtel, A., Klamroth, J., & Tichy, W. F. (2017, March). Natural language user interface for software engineering tasks. In *Proceedings of the International Conference on Advances in Computer-Human Interactions (ACHI)* (*Vol. 10*, pp. 34-39). Academic Press.

Xu, C., Pathak, P. H., & Mohapatra, P. (2015, February). Finger-writing with smartwatch: A case for finger and hand gesture recognition using smartwatch. In *Proceedings of the 16th International Workshop on Mobile Computing Systems and Applications* (pp. 9-14). 10.1145/2699343.2699350

Compilation of References

Aggarwal, J. K., & Xia, L. (2014). Human activity recognition from 3d data: A review. *Pattern Recognition Letters*, *48*, 70–80. doi:10.1016/j.patrec.2014.04.011

Alani. (2018). Hand Gesture Recognition Using an Adapted Convolutional Neural Network with Data Augmentation. *24th IEEE International Conference on Information Management*.

Albiol, A., Torres, L., & Delp, E. J. (2001). Optimum color spaces for skin detection. *Proceedings - International Conference on Image Processing*, *1*, 122–124.

Attwenger, A. (2017). *Advantages and Drawbacks of Gesture-Based Interaction*. Grin Verlag.

Banjarey, K., Prakash Sahu, S., & Kumar Dewangan, D. (2021). A Survey on Human Activity Recognition using Sensors and Deep Learning Methods. *2021 5th International Conference on Computing Methodologies and Communication (ICCMC)*, 1610-1617. 10.1109/ICCMC51019.2021.9418255

Banjarey, K., Sahu, S. P., & Dewangan, D. K. (2022). Human Activity Recognition Using 1D Convolutional Neural Network. In S. Shakya, V. E. Balas, S. Kamolphiwong, & K. L. Du (Eds.), *Sentimental Analysis and Deep Learning. Advances in Intelligent Systems and Computing* (Vol. 1408). Springer. doi:10.1007/978-981-16-5157-1_54

Basak, S., & Chowdhury, A. (2014). A vision interface framework for intuitive gesture recognition using color based blob detection. *International Journal of Computers and Applications*, *90*(15).

Baum, L. (1972). An inequality and associated maximization technique in statistical estimation of probabilistic functions of markov processes. *Inequalities*, *3*, 1–8.

Baum, L., Petrie, T., Soules, G., & Weiss, N. (1970). A maximization technique occurring in thestatistical analysis of probabilistic functions of markov chains. *Annals of Mathematical Statistics*, *41*(1), 164171. doi:10.1214/aoms/1177697196

Bhushan, A., Rastogi, P., Sharma, H. K., & Ahmed, M. E. (2017). I/O and memory management: Two keys for tuning RDBMS. *Proceedings on 2016 2nd International Conference on Next Generation Computing Technologies, NGCT 2016*, 208–214.

Bhuyan, M. K., Neog, D. R., & Kar, M. K. (2012). Fingertip detection for hand pose recognition. *International Journal on Computer Science and Engineering*, *4*(3), 501.

Birchfield, S. (1998). Elliptical head tracking using intensity gradients and color histograms. *Proceedings of CVPR '98*, 232–237. 10.1109/CVPR.1998.698614

Bobick, A., & Wilson, A. (1995). Using configuration states for the representation and recognition of gesture. In *Proc. Fifth International Conf. on Computer Vision*. IEEE Press.

Brand, J., & Mason, J. (2000). A comparative assessment of three approaches to pixel evel human skin-detection. *Proc. of the International Conference on Pattern Recognition*, 1, 1056–1059. 10.1109/ICPR.2000.905653

Bregler, C., & Omohundro, S. (1995). Nonlinear manifold learning for visual speech recogni-tion. In *Proc. Fifth International Conf. on Computer Vision*. IEEE Press. 10.1109/ICCV.1995.466899

Brown, D., Craw, I., & Lewthwaite, J. (2001). A som based approach to skin detection with application in real time systems. *Proc. of the British Machine Vision Conference*. 10.5244/C.15.51

Challa, S. K., Kumar, A., & Semwal, V. B. (2021). A multibranch CNN-BiLSTM model for human activity recognition using wearable sensor data. *The Visual Computer*. Advance online publication. doi:10.100700371-021-02283-3

Chen, C., Jafari, R., & Kehtarnavaz, N. (2015, January). Action recognition from depth sequences using depth motion maps-based local binary patterns. In *2015 IEEE Winter Conference on Applications of Computer Vision* (pp. 1092-1099). IEEE. 10.1109/WACV.2015.150

Chen, C., Jafari, R., & Kehtarnavaz, N. (2017). A survey of depth and inertial sensor fusion for human action recognition. *Multimedia Tools and Applications*, *76*(3), 4405–4425. doi:10.100711042-015-3177-1

Chen, C., Liu, K., & Kehtarnavaz, N. (2016). Real-time human action recognition based on depth motion maps. *Journal of Real-Time Image Processing*, *12*(1), 155–163. doi:10.100711554-013-0370-1

Chen, L., Wang, F., Deng, H., & Ji, K. (2013, December). A survey on hand gesture recognition. In *2013 International conference on computer sciences and applications* (pp. 313-316). IEEE. 10.1109/CSA.2013.79

Chen, L., Wei, H., & Ferryman, J. (2013). A survey of human motion analysis using depth imagery. *Pattern Recognition Letters*, *34*(15), 1995–2006. doi:10.1016/j.patrec.2013.02.006

Chen, Q., Wu, H., & Yachida, M. (1995). Face detection by fuzzy pattern matching. *Proc. of the Fifth International Conference on Computer Vision*, 591–597. 10.1109/ICCV.1995.466885

Chu, S., & Tanaka, J. (2011, July). Hand gesture for taking self-portrait. In *International Conference on Human-Computer Interaction* (pp. 238-247). Springer.

Clough, S., & Duff, M. C. (2020). The role of gesture in communication and cognition: Implications for understanding and treating neurogenic communication disorders. *Frontiers in Human Neuroscience*, *14*, 323. doi:10.3389/fnhum.2020.00323 PMID:32903691

Cui, Y., Swets, D., & Weng, J. (1995). Learning-based hand sign recognition using SHOSH-LIF-M. In *Proc. Fifth International Conf. on Computer Vision*. IEEE Press.

Dai, D., Zhuang, W., Shen, Y., Li, L., & Wang, H. (2020, July). Design of Intelligent Mobile Robot Control System Based on Gesture Recognition. In *International Conference on Artificial Intelligence and Security* (pp. 101-111). Springer. 10.1007/978-981-15-8086-4_10

Dardas, N. H., & Georganas, N. D. (2011). Real-time hand gesture detection and recognition using bag-of-features and support vector machine techniques. *IEEE Transactions on Instrumentation and Measurement*, *60*(11), 3592–3607. doi:10.1109/TIM.2011.2161140

Dardas, N., Chen, Q., Georganas, N. D., & Petriu, E. M. (2010, October). Hand gesture recognition using bag-of-features and multi-class support vector machine. In *2010 IEEE International Symposium on Haptic Audio Visual Environments and Games* (pp. 1-5). IEEE. 10.1109/HAVE.2010.5623982

Devi & Thakur. (2017). Content Aware Video Compression:-An Approach To VOS Algorithm. *International Journal of Trend in Research and Development, 4*(4).

Dewangan & Sahu. (2021). Driving Behavior Analysis of Intelligent Vehicle System for Lane Detection Using Vision-Sensor. *IEEE Sensors Journal, 21*(5), 6367-6375. . doi:10.1109/JSEN.2020.3037340

Dinh, D. L., Kim, J. T., & Kim, T. S. (2014). Hand gesture recognition and interface via a depth imaging sensor for smart home appliances. *Energy Procedia*, *62*, 576–582. doi:10.1016/j.egypro.2014.12.419

Dipietro, L., Sabatini, A. M., & Dario, P. (2008). A survey of glove-based systems and their applications. *IEEE Transactions on Systems, Man, and Cybernetics, Part C (Applications and Reviews), 38*(4), 461-482.

Dominio, F., Donadeo, M., & Zanuttigh, P. (2014). Combining multiple depth-based descriptors for hand gesture recognition. *Pattern Recognition Letters*, *50*, 101–111. doi:10.1016/j.patrec.2013.10.010

Dong, Y., Liu, J., & Yan, W. (2021). Dynamic Hand Gesture Recognition Based on Signals From Specialized Data Glove and Deep Learning Algorithms. *IEEE Transactions on Instrumentation and Measurement*, *70*, 1–14. doi:10.1109/TIM.2021.3077967

Dua, N., Singh, S. N., & Semwal, V. B. (2021). Multi-input CNN-GRU based human activity recognition using wearable sensors. *Computing*, *103*(7), 1461–1478. doi:10.100700607-021-00928-8

Duda, R., & Hart, P. (1973). *Pattern Classification and Scene Analysis*. John Wiley & Sons, Inc.

Elmezain, M., Al-Hamadi, A., Sadek, S., & Michaelis, B. (2010, December). Robust methods for hand gesture spotting and recognition using hidden markov models and conditional random fields. In *The 10th IEEE International Symposium on Signal Processing and Information Technology* (pp. 131-136). IEEE. 10.1109/ISSPIT.2010.5711749

Elmezain, M., Al-Hamadi, A., Krell, G., El-Etriby, S., & Michaelis, B. (2007, December). *Gesture recognition for alphabets from hand motion trajectory using hidden Markov models. In 2007 IEEE International Symposium on Signal Processing and Information Technology.* IEEE.

Erol, A., Bebis, G., Nicolescu, M., Boyle, R. D., & Twombly, X. (2007). Vision-based hand pose estimation: A review. *Computer Vision and Image Understanding, 108*(1-2), 52–73. doi:10.1016/j. cviu.2006.10.012

Evangelidis, G., Singh, G., & Horaud, R. (2014, August). Skeletal quads: Human action recognition using joint quadruples. In *2014 22nd International Conference on Pattern Recognition* (pp. 4513-4518). IEEE.

Fang, Y., Cheng, J., Wang, K., & Lu, H. (2007, August). Hand gesture recognition using fast multi-scale analysis. In *Fourth international conference on image and graphics (ICIG 2007)* (pp. 694-698). IEEE. 10.1109/ICIG.2007.52

Gałka, J., Mąsior, M., Zaborski, M., & Barczewska, K. (2016). Inertial motion sensing glove for sign language gesture acquisition and recognition. *IEEE Sensors Journal, 16*(16), 6310–6316. doi:10.1109/JSEN.2016.2583542

Gao, Q., Ogenyi, U. E., Liu, J., Ju, Z., & Liu, H. (2019, September). A two-stream CNN framework for American sign language recognition based on multimodal data fusion. In *UK Workshop on Computational Intelligence* (pp. 107-118). Springer.

Geetha, M., & Manjusha, U. C. (2012). A vision based recognition of indian sign language alphabets and numerals using b-spline approximation. *International Journal on Computer Science and Engineering, 4*(3), 406.

Gomez, G. (2000). On selecting colour components for skin detection. *Proc. of the ICPR, 2,* 961–964.

Gomez, G., & Morales, E. (2002). Automatic feature construction and a simple rule induction algorithm for skin detection. *Proc. of the ICML Workshop on Machine Learning in Computer Vision,* 31–38.

Grandini, M., Bagli, E., & Visani, G. (2020). *Metrics for multi-class classification: an overview.* arXiv preprint arXiv:2008.05756.

Gunawardane, P. D. S. H., & Medagedara, N. T. (2017, October). Comparison of hand gesture inputs of leap motion controller & data glove into a soft finger. In *2017 IEEE International Symposium on Robotics and Intelligent Sensors (IRIS)* (pp. 62-68). IEEE. 10.1109/IRIS.2017.8250099

Guo, J., & Li, S. (2011). *Hand gesture recognition and interaction with 3D stereo camera.* The Project Report of Australian National University.

Gupta, R., Gupta, A., & Aswal, R. (2021, January). Time-CNN and Stacked LSTM for Posture Classification. In *2021 International Conference on Computer Communication and Informatics (ICCCI)* (pp. 1-5). IEEE. 10.1109/ICCCI50826.2021.9402657

Gupta, R., & Kumar, A. (2021). Indian sign language recognition using wearable sensors and multi-label classification. *Computers & Electrical Engineering, 90,* 106898. doi:10.1016/j.compeleceng.2020.106898

Hafiz, M. A.-R., Lehmia, K., Danish, M., & Noman, M. (2017). CMSWVHG-control MS Windows via hand gesture. In *International Multi-Topic Conference INMIC,* 1-7.

Han, J., Shao, L., Xu, D., & Shotton, J. (2013). Enhanced computer vision with microsoft kinect sensor: A review. *IEEE Transactions on Cybernetics, 43*(5), 1318–1334. doi:10.1109/TCYB.2013.2265378 PMID:23807480

Haria, A., Subramanian, A., Asokkumar, N., Poddar, S., & Nayak, J. S. (2017). Hand gesture recognition for human computer interaction. *Procedia Computer Science, 115,* 367–374. doi:10.1016/j.procs.2017.09.092

Hasan, M. M., & Mishra, P. K. (2012). Hand gesture modeling and recognition using geometric features: a review. *Canadian Journal on Image Processing and Computer Vision, 3*(1), 12-26.

Hema, D., & Kannan, D. S. (2020). *Interactive color image segmentation using HSV color space.* Science and Technology Journal.

Hexa, R. (2017). Gesture recognition market analysis by technology 2D. 3D. By application tablets & notebooks. Smartphones. Gaming consoles. Smart televisions. In Laptops & desktops and segment forecasts 2014-2024. IEEE.

Huang, X., Ariki, Y., & Jack, M. (1990). *Hidden Markov Models for Speech Recognition.* Edinburgh University Press.

Ibrahim, A., Hashim, A., Faisal, K., & Youngwook, K. (2018). Hand gesture recognition using input impedance variation of two antennas with transfer learning. *Sensors Journal, 18,* 4129-4135.

Jalab, H. A. (2012). Static hand Gesture recognition for human computer interaction. *Information Technology Journal, 11*(9), 1265–1271. doi:10.3923/itj.2012.1265.1271

Jeong, M. H., Kuno, Y., Shimada, N., & Shirai, Y. (2002, August). Two-hand gesture recognition using coupled switching linear model. In *Object recognition supported by user interaction for service robots* (Vol. 3, pp. 529–532). IEEE.

Jones, M. J., & Rehg, J. M. (1999). Statistical color models with application to skin detection. *Proc. of the CVPR '99, 1,* 274–280. 10.1109/CVPR.1999.786951

Kaâniche, M. (2009). *Gesture recognition from video sequences* (Doctoral dissertation). Université Nice Sophia Antipolis.

Kane, L., & Khanna, P. (2015). A framework for live and cross platform fingerspelling recognition using modified shape matrix variants on depth silhouettes. *Computer Vision and Image Understanding, 141,* 138–151. doi:10.1016/j.cviu.2015.08.001

Kane, L., & Khanna, P. (2019). Depth matrix and adaptive Bayes classifier based dynamic hand gesture recognition. *Pattern Recognition Letters*, *120*, 24–30. doi:10.1016/j.patrec.2019.01.003

Karthik, S. K., Akash, S., Srinath, R., & Shitij, K. (2017). Recognition of human arm gestures using Myo armband for the game of hand cricket. *International symposium on robotics and intelligent sensors IRIS*, 389-394.

Kasprzak, W., Wilkowski, A., & Czapnik, K. (2012). Hand gesture recognition based on free-form contours and probabilistic inference. *International Journal of Applied Mathematics and Computer Science*, *22*, 437–448.

Khomami, S. A., & Shamekhi, S. (2021). Persian sign language recognition using IMU and surface EMG sensors. *Measurement*, *168*, 108471. doi:10.1016/j.measurement.2020.108471

Kim, J. B., Park, K. H., Bang, W. C., & Bien, Z. Z. (2002, May). Continuous gesture recognition system for Korean sign language based on fuzzy logic and hidden Markov model. In *2002 IEEE World Congress on Computational Intelligence. 2002 IEEE International Conference on Fuzzy Systems. FUZZ-IEEE'02. Proceedings (Cat. No. 02CH37291)* (Vol. 2, pp. 1574-1579). IEEE.

Kim, J. H., Thang, N. D., & Kim, T. S. (2009, July). 3-D hand motion tracking and gesture recognition using a data glove. In *2009 IEEE International Symposium on Industrial Electronics* (pp. 1013-1018). IEEE. 10.1109/ISIE.2009.5221998

Kim, L., & Park. (2008). Dynamic hand gesture recognition using a CNN model with 3D receptive fields. *International Conference on Neural Networks and Signal Processing*.

Kolivand, H., Joudaki, S., Sunar, M. S., & Tully, D. (2021). A new framework for sign language alphabet hand posture recognition using geometrical features through artificial neural network (part 1). *Neural Computing & Applications*, *33*(10), 4945–4963. doi:10.100700521-020-05279-7

Kraljević, L., Russo, M., Pauković, M., & Šarić, M. (2020). A Dynamic Gesture Recognition Interface for Smart Home Control based on Croatian Sign Language. *Applied Sciences (Basel, Switzerland)*, *10*(7), 2300. doi:10.3390/app10072300

Kreuzer, D., & Munz, M. (2021). Deep Convolutional and LSTM Networks on Multi-Channel Time Series Data for Gait Phase Recognition. *Sensors (Basel)*, *21*(3), 789. doi:10.339021030789 PMID:33503947

Krizhevsky, Sutskever, & Hinton. (2012). *Imagenet classification with deep convolutional neural networks*. NIPS.

Kudrinko, K., Flavin, E., Zhu, X., & Li, Q. (2020). Wearable Sensor-Based Sign Language Recognition: A Comprehensive Review. *IEEE Reviews in Biomedical Engineering*, *14*, 82–97. doi:10.1109/RBME.2020.3019769 PMID:32845843

Kumar Sharma, H., & Kshitiz, K. (2018). NLP and Machine Learning Techniques for Detecting Insulting Comments on Social Networking Platforms. *ICACCE 2018*.

Kumar Sharma, H., & Kshitiz, K. (2018). *NLP and Machine Learning Techniques for Detecting Insulting Comments on Social Networking Platforms*. ICACCE.

Kumarage, D., Fernando, S., Fernando, P., Madushanka, D., & Samarasinghe, R. (2011, August). Real-time sign language gesture recognition using still-image comparison & motion recognition. In *2011 6th International Conference on Industrial and Information Systems* (pp. 169-174). IEEE.

Kundu, A. S., Mazumder, O., Lenka, P. K., & Bhaumik, S. (2018). Hand gesture recognition based omnidirectional wheelchair control using IMU and EMG sensors. *Journal of Intelligent & Robotic Systems*, *91*(3), 529–541. doi:10.100710846-017-0725-0

Lahamy, H., & Lichti, D. (2012). Robust Real-Time and Rotation-Invariant American Sign Language Alphabet Recognition Using Range Camera. *Proceedings of the International Archives of the Photogrammetry, Remote Sensing and Spatial Information Sciences*, 217-222.

Lamar, M. V., Bhuiyan, M. S., & Iwata, A. (1999, October). Hand gesture recognition using morphological principal component analysis and an improved CombNET-II. In *IEEE SMC'99 Conference Proceedings. 1999 IEEE International Conference on Systems, Man, and Cybernetics (Cat. No. 99CH37028)* (Vol. 4, pp. 57-62). IEEE.

Lee, J. Y., & Yoo, S. I. (2002). An elliptical boundary model for skin color detection. *Proc. of the 2002 International Conference on Imaging Science, Systems, and Technology*.

Lee, Y. H., Ahn, H., Ahn, H. B., & Lee, S. Y. (2019). Visual object detection and tracking using analytical learning approach of validity level. *Intelligent Automation and Soft Computing*, *25*(1), 205–215.

Liang, R. H., & Ouhyoung, M. (1998, April). A real-time continuous gesture recognition system for sign language. In *Proceedings third IEEE international conference on automatic face and gesture recognition* (pp. 558-567). IEEE.

Li, W., Zhang, Z., & Liu, Z. (2010, June). Action recognition based on a bag of 3d points. In *2010 IEEE Computer Society Conference on Computer Vision and Pattern Recognition-Workshops* (pp. 9-14). IEEE.

Li, Y. (2012, June). Hand gesture recognition using Kinect. In *2012 IEEE International Conference on Computer Science and Automation Engineering* (pp. 196-199). IEEE.

Li, Y., Ma, D., Yu, Y., Wei, G., & Zhou, Y. (2021). Compact joints encoding for skeleton-based dynamic hand gesture recognition. *Computers & Graphics*, *97*, 191–199.

Madhuri, Y., Anitha, G., & Anburajan, M. (2013, February). Vision-based sign language translation device. In *2013 International Conference on Information Communication and Embedded Systems (ICICES)* (pp. 565-568). IEEE.

Maini, R., & Aggarwal, H. (2009). Study and comparison of various image edge detection techniques. *International Journal of Image Processing (IJIP)*, *3*(1), 1-11.

Mantecón, T., del Blanco, C. R., Jaureguizar, F., & García, N. (2016). Hand Gesture Recognition using Infrared Imagery Provided by Leap Motion Controller. *Int. Conf. on Advanced Concepts for Intelligent Vision Systems, ACIVS 2016*, 47-57. 10.1007/978-3-319-48680-2_5

Marco, E. B., Andrés, G. J., Jonathan, A. Z., Andrés, P., & Víctor, H. A. (2017a). Hand gesture recognition using machine learning and the Myo armband. In *European Signal Processing Conference*. IEEE.

Marco, E. B., Cristhian, M., Jonathan, A. Z., Andrés, G. J., Carlos, E. A., Patricio, Z., Marco, S., Freddy, B. P., & María, P. (2017b). Real-time hand gesture recognition using the Myo armband and muscle activity detection. In *Second Ecuador technical chapters meeting ETCM*. IEEE.

Masood, S., Srivastava, A., Thuwal, H. C., & Ahmad, M. (2018). Real-time sign language gesture (word) recognition from video sequences using CNN and RNN. In *Intelligent Engineering Informatics* (pp. 623–632). Springer. doi:10.1007/978-981-10-7566-7_63

Meena, S. (2011). *A study on hand gesture recognition technique* (Doctoral dissertation).

Min, B. W., Yoon, H. S., Soh, J., Yang, Y. M., & Ejima, T. (1997, October). Hand gesture recognition using hidden Markov models. In *1997 IEEE International Conference on Systems, Man, and Cybernetics. Computational Cybernetics and Simulation* (Vol. 5, pp. 4232-4235). IEEE.

Mitra, S., & Acharya, T. (2007). *Gesture recognition: a survey. IEEE Trans Syst Man Cybern (SMC) Part C Apple.*

Muezzinoglu, T., & Karakose, M. (2021). An Intelligent Human–Unmanned Aerial Vehicle Interaction Approach in Real Time Based on Machine Learning Using Wearable Gloves. *Sensors (Basel)*, *21*(5), 1766. doi:10.339021051766 PMID:33806388

Murthy, G. R. S., & Jadon, R. S. (2009). A review of vision based hand gestures recognition. *International Journal of Information Technology and Knowledge Management*, *2*(2), 405–410.

Nag, R., Wong, K., & Fallside, F. (1986). Script recognition using hidden Markov models. ICASSP 86. doi:10.1109/ICASSP.1986.1168951

Nagarajan, S., & Subashini, T. S. (2013). Static hand gesture recognition for sign language alphabets using edge oriented histogram and multi class SVM. *International Journal of Computers and Applications*, *82*(4).

Nishikawa, A., Hosoi, T., Koara, K., Negoro, D., Hikita, A., Asano, S., Kakutani, H., Miyazaki, F., Sekimoto, M., Yasui, M., Miyake, Y., Takiguchi, S., & Monden, M. (2003). FAce MOUSe: A novel human-machine interface for controlling the position of a laparoscope. *IEEE Transactions on Robotics and Automation*, *19*(5), 825–841.

Oka, K., Sato, Y., & Koike, H. (2002, May). Real-time tracking of multiple fingertips and gesture recognition for augmented desk interface systems. In *Proceedings of Fifth IEEE International Conference on Automatic Face Gesture Recognition* (pp. 429-434). IEEE.

Oreifej, O., & Liu, Z. (2013). Hon4d: Histogram of oriented 4d normals for activity recognition from depth sequences. In *Proceedings of the IEEE conference on computer vision and pattern recognition* (pp. 716-723). IEEE.

Panwar, M. (2012, February). Hand gesture recognition based on shape parameters. In *2012 International Conference on Computing, Communication and Applications* (pp. 1-6). IEEE.

Pardhi, P., Yadav, K., Shrivastav, S., Sahu, S. P., & Kumar Dewangan, D. (2021). Vehicle Motion Prediction for Autonomous Navigation system Using 3 Dimensional Convolutional Neural Network. *2021 5th International Conference on Computing Methodologies and Communication (ICCMC)*, 1322-1329. 10.1109/ICCMC51019.2021.9418449

Park, J. J., & Kwon, C. K. (2021). Korean Finger Number Gesture Recognition Based on CNN Using Surface Electromyography Signals. *Journal of Electrical Engineering & Technology*, *16*(1), 591–598. doi:10.100742835-020-00587-3

Patni, J. C., & Sharma, H. K. (2019). Air Quality Prediction using Artificial Neural Networks. *2019 International Conference on Automation, Computational and Technology Management, ICACTM 2019*, 568–572. 10.1109/ICACTM.2019.8776774

Piotr, K., Tomasz, M., & Jakub, T. (2017). Towards sensor position-invariant hand gesture recognition using a mechanomyographic interface. In Signal processing: algorithms, architectures, arrangements and applications SPA. IEEE.

Pradhan, A., Ghose, M. K., Pradhan, M., Qazi, S., & Moors, T., El-Arab, I. M. E., ... Memon, A. (2012). A hand gesture recognition using feature extraction. *Int J Curr Eng Technol*, *2*(4), 323–327.

Prashan, P. (2014). Historical development of hand gesture recognition. In Cognitive science and technology book series CSAT. Singapore: Springer.

Praveen, K.S., & Shreya, S. (2015). Evolution of hand gesture recognition: A review. *International Journal of Engineering and Computer Science, 4*, 9962-9965.

Rabiner, L., & Juang, B. (1986). An introduction to hidden markov models. *IEEE ASSP Magazine*, *3*(January), 4–16. doi:10.1109/MASSP.1986.1165342

Rahmat, R. F., Chairunnisa, T. E. N. G. K. U., Gunawan, D. A. N. I., Pasha, M. F., & Budiarto, R. A. H. M. A. T. (2019). Hand gestures recognition with improved skin color segmentation in human-computer interaction applications. *Journal of Theoretical and Applied Information Technology*, *97*(3), 727–739.

Rajesh, R. J., Nagarjunan, D., Arunachalam, R. M., & Aarthi, R. (2012). Distance Transform Based Hand Gestures Recognition for PowerPoint Presentation Navigation. *Advances in Computers*, *3*(3), 41.

Ramakrishna mission Vivekananda University, Coimbatore Campus. (2021). *Indian Sign Language Dictionary*. http://indiansignlanguage.org

Ramjan, M. R., Sandip, R. M., Uttam, P. S., & Srimant, W. S. (2014). Dynamic hand gesture recognition and detection for real time using human computer interaction. *International Journal (Toronto, Ont.)*, 2(3).

Rastgoo, R., Kiani, K., & Escalera, S. (2020). Sign language recognition: A deep survey. *Expert Systems with Applications*, 113794.

Rautaray, S. S., & Agrawal, A. (2015). Vision based hand gesture recognition for human computer interaction: A survey. *Artificial Intelligence Review*, 43(1), 1–54. doi:10.100710462-012-9356-9

Rekha, J., Bhattacharya, J., & Majumder, S. (2011). Hand gesture recognition for sign language: A new hybrid approach. In *Proceedings of the International Conference on Image Processing, Computer Vision, and Pattern Recognition (IPCV)* (p. 1). The Steering Committee of The World Congress in Computer Science, Computer Engineering and Applied Computing (WorldComp).

Rekha, J., Bhattacharya, J., & Majumder, S. (2011, December). Shape, texture and local movement hand gesture features for indian sign language recognition. In *3rd International Conference on Trendz in Information Sciences & Computing (TISC2011)* (pp. 30-35). IEEE.

Ren, Z., Yuan, J., Meng, J., & Zhang, Z. (2013). Robust part-based hand gesture recognition using kinect sensor. *IEEE Transactions on Multimedia*, 15(5), 1110–1120. doi:10.1109/TMM.2013.2246148

Rishabh, S., Raj, S., Nutan, V. B., & Prachi, R. R. (2016). Interactive projector screen with hand detection using gestures. In *International conference on automatic control and dynamic optimization techniques ICACDOT*. IEEE.

Samata, M., & Kinage, K.S. (2015). Study on hand gesture recognition. *International Journal of Computer Science and Mobile Computing, 4*, 51-57.

Schultz, M., Gill, J., Zubairi, S., Huber, R., & Gordin, F. (2003). Bacterial contamination of computer keyboards in a teaching hospital. *Infection Control and Hospital Epidemiology*, 4(24), 302–303.

Schumeyer, R., & Barner, K. (1998). A color-based classifier for region identification in video. In *Visual Communications and Image Processing 1998* (Vol. 3309, pp. 189–200). SPIE.

Sharma, S., Gupta, R., & Kumar, A. (2019, March). On the use of multi-modal sensing in sign language classification. In *2019 6th International Conference on Signal Processing and Integrated Networks (SPIN)* (pp. 495-500). IEEE. 10.1109/SPIN.2019.8711702

Sharma, Shastri, & Biswas. (2013). A Framework for Automated Database Tuning Using Dynamic SGA Parameters and Basic Operating System Utilities. *Database System Journal*.

Sharma, Shastri, & Biswas. (2015). Auto-selection and management of dynamic SGA parameters in RDBMS. *Database System Journal*.

Shin, M. C., Chang, K. I., & Tsap, L. V. (2002). Does colorspace transformation make any difference on skin detection? *IEEE Workshop on Applications of Computer Vision*. 10.1109/ACV.2002.1182194

Sigal, L., Sclaroff, S., & Athitsos, V. (2000). Estimation and prediction of evolving color distributions for skin segmentation under varying illumination. *Proc. IEEE Conf. on Computer Vision and Pattern Recognition, 2*, 152–159. 10.1109/CVPR.2000.854764

Siji Rani, S., Dhrisya, K. J., & Ahalyadas, M. (2017). International conference on advances in computing. In *Communications and informatics ICACCI*. IEEE.

Soriano, M., Huovinen, S., Martinkauppi, B., & Laaksonen, M. (2000). Skin detection in video under changing illumination conditions. *Proc. 15th International Conference on Pattern Recognition, 1*, 839–842. 10.1109/ICPR.2000.905542

Starner, T., Makhoul, J., Schwartz, R., & Chou, G. (1994). On-line cursive hand-writing recognition using speech recognition methods. ICASSP 94.

Starner, T., & Pentland, A. (1995). Visual recognition of American Sign Language using hiddenmarkov models. *Proc. of the Intl. Workshop on Automatic Face- and Gesture-Recognition.*

Stefano, S., Paolo, M. R., David, A. F. G., Fabio, R., Rossana, T., Elisa, C., & Danilo, D. (2018). On-line event-driven hand gesture recognition based on surface electromyographic signals. In *International symposium on circuits and systems ISCAS*. IEEE.

Stergiopoulou, E., & Papamarkos, N. (2009). Hand gesture recognition using a neural network shape fitting technique. *Engineering Applications of Artificial Intelligence, 22*(8), 1141–1158.

Stern, H., & Efros, B. (2002). Adaptive color space switching for face tracking in multi-colored lighting environments. *Proc. of the International Conference on Automatic Face and Gesture Recognition*, 249–255. 10.1109/AFGR.2002.1004162

Sultana, A., & Rajapuspha, T. (2012). Vision based gesture recognition for alphabetical hand gestures using the svm classifier. *International Journal of Computer Science and Engineering Technology, 3*(7), 218–223.

Suri, K., & Gupta, R. (2019). Continuous sign language recognition from wearable IMUs using deep capsule networks and game theory. *Computers & Electrical Engineering, 78*, 493–503. doi:10.1016/j.compeleceng.2019.08.006

Szegedy, L., Jia, S., Reed, A., & Erhan, V., & Rabinovich. (2015). Going deeper with convolutions. *2015 IEEE Conference on Computer Vision and Pattern Recognition (CVPR).*

Taneja, Karthik, Shukla, & Sharma. (n.d.). Architecture of IOT based Real Time Tracking System. *International Journal of Innovations & Advancement in Computer Science, 6*(12).

Taneja, S., Karthik, M., Shukla, M., & Sharma, H. K. (2017). *AirBits: A Web Application Development using Microsoft Azure*. International Conference on Recent Developments in Science, Technology, Humanities and Management (ICRDSTHM-17), Kuala Lumpur, Malasyia.

Taneja, S., Karthik, M., Shukla, M., & Sharma, H. K. (2017). *AirBits: A Web Application Development using Microsoft Azure.* International Conference on Recent Developments in Science, Technology, Humanities and Management (ICRDSTHM-17), Kuala Lumpur, Malaysia.

Taneja, S., Karthik, M., Shukla, M., & Sharma, H. K. (2018). Architecture of IOT based Real Time Tracking System. *International Journal of Innovations & Advancement in Computer Science, 6*(12).

Terrillon, J.-C., Shirazi, M. N., Fukamachi, H., & Akamatsu, S. (2000). Comparative performance of different skin chrominance models and chrominance spaces for the automatic detection of human faces in color images. *Proc. of the International Conference on Face and Gesture Recognition,* 54–61. 10.1109/AFGR.2000.840612

Tewari, A., Taetz, B., Grandidier, F., & Stricker, D. (2017, October). A probabilistic combination of CNN and RNN estimates for hand gesture based interaction in car. In *2017 IEEE International Symposium on Mixed and Augmented Reality (ISMAR-Adjunct)* (pp. 1-6). IEEE.

Vaibhavi, S.G., Akshay, A.K., Sanket, N.R., Vaishali, A.T., & Shabnam, S.S. (2014). A review of various gesture recognition techniques. *International Journal of Engineering and Computer Science, 3*, 8202-8206.

Vezhnevets, V., Sazonov, V., & Andreeva, A. (2003). *A Survey on Pixel-Based Skin Color Detection Techniques.* In International Conference Graphicon 2003, Moscow, Russia.

Vieira, A. W., Nascimento, E. R., Oliveira, G. L., Liu, Z., & Campos, M. F. (2012, September). Stop: Space-time occupancy patterns for 3d action recognition from depth map sequences. In *Iberoamerican congress on pattern recognition* (pp. 252–259). Springer.

Wachtel, A., Klamroth, J., & Tichy, W. F. (2017, March). Natural language user interface for software engineering tasks. In *Proceedings of the International Conference on Advances in Computer-Human Interactions (ACHI)* (Vol. 10, pp. 34-39). Academic Press.

Wadhawan, A., & Kumar, P. (2020). Deep learning-based sign language recognition system for static signs. *Neural Computing & Applications, 32*(12), 7957–7968. doi:10.100700521-019-04691-y

Wadhawan, A., & Kumar, P. (2021). Sign language recognition systems: A decade systematic literature review. *Archives of Computational Methods in Engineering, 28*(3), 785–813. doi:10.100711831-019-09384-2

Wang, J., Liu, Z., Chorowski, J., Chen, Z., & Wu, Y. (2012, October). Robust 3d action recognition with random occupancy patterns. In *European Conference on Computer Vision* (pp. 872-885). Springer.

Wilson, A., & Bobick, A. (1995). Learning Visual Behavior for Gesture Analysis. *Proc. IEEE Symposium on Computer Vision.*

Wu, Y., & Huang, T. S. (2001). Hand modeling, analysis and recognition. *IEEE Signal Processing Magazine, 18*(3), 51–60.

Xia, L., & Aggarwal, J. K. (2013). Spatio-temporal depth cuboid similarity feature for activity recognition using depth camera. In *Proceedings of the IEEE conference on computer vision and pattern recognition* (pp. 2834-2841). IEEE.

Xu, C., Pathak, P. H., & Mohapatra, P. (2015, February). Finger-writing with smartwatch: A case for finger and hand gesture recognition using smartwatch. In *Proceedings of the 16th International Workshop on Mobile Computing Systems and Applications* (pp. 9-14). 10.1145/2699343.2699350

Yamato, J., Ohya, J., & Ishii, K. (1992). Recognizing human action in time-sequential images using hidden markov models. In *Proc. 1992 IEEE Conf. on Computer Vision and Pattern Recognition*. IEEE Press. 10.1109/CVPR.1992.223161

Yang, Z., Li, Y., Chen, W., & Zheng, Y. (2012, July). Dynamic hand gesture recognition using hidden Markov models. In *2012 7th International Conference on Computer Science & Education (ICCSE)* (pp. 360-365). IEEE.

Yang, M., & Ahuja, N. (1999). Gaussian mixture model for human skin color and its application in image and video databases. *Proc. of the SPIE: Conf. on Storage and Retrieval for Image and Video Databases (SPIE 99)*, 3656, 458–466 10.1117/12.333865

Yang, X., Zhang, C., & Tian, Y. (2012, October). Recognizing actions using depth motion maps-based histograms of oriented gradients. In *Proceedings of the 20th ACM international conference on Multimedia* (pp. 1057-1060). ACM.

Yasen, M., & Jusoh, S. (2019, September). A systematic review on hand gesture recognition techniques, challenges and applications. *PeerJ. Computer Science*, 5, e218. doi:10.7717/peerj-cs.218 PMID:33816871

Ye, M., Zhang, Q., Wang, L., Zhu, J., Yang, R., & Gall, J. (2013). *A survey on human motion analysis from depth data*. Academic Press.

Yifan, Z., Congqi, C., Jian, C., & Hanqing, L. (2018). EgoGesture: A new dataset and benchmark for egocentric hand gesture recognition. *Transactions on Multimedia, 20*, 1038-1050. DOI . doi:10.1109/TMM.2018.2808769

Yiming, Z., & Liuai, W. (2021). Human Motion State Recognition Based on Multi-input ConvLSTM. *Journal of Physics: Conference Series, 1748*(6), 062075. doi:10.1088/1742-6596/1748/6/062075

Yoon, H. S., Soh, J., Bae, Y. J., & Yang, H. S. (2001). Hand gesture recognition using combined features of location, angle and velocity. *Pattern Recognition, 34*(7), 1491–1501.

Yuan, G., Liu, X., Yan, Q., Qiao, S., Wang, Z., & Yuan, L. (2020). Hand gesture recognition using deep feature fusion network based on wearable sensors. *IEEE Sensors Journal, 21*(1), 539–547. doi:10.1109/JSEN.2020.3014276

Yuntao, M., Yuxuan, L., Ruiyang, J., Xingyang, Y., Raza, S., Samuel, W., & Ravi, V. (2017). *Hand gesture recognition with convolutional neural networks for the multimodal UAV control*. IEEE.

Zandigohar, M., Han, M., Sharif, M., Gunay, S. Y., Furmanek, M. P., Yarossi, M., . . . Schirner, G. (2021). *Multimodal Fusion of EMG and Vision for Human Grasp Intent Inference in Prosthetic Hand Control.* arXiv preprint arXiv:2104.03893.

Zarit, B. D., Super, B. J., & Quek, F. K. H. (1999). Comparison of five color models in skin pixel classification. *ICCV'99 Int'l Workshop on recognition, analysis and tracking of faces and gestures in Real-Time systems*, 58–63. 10.1109/RATFG.1999.799224

Zengeler, N., Kopinski, T., & Handmann, U. (2019). Hand Gesture Recognition in Automotive Human–Machine Interaction Using Depth Cameras. *Sensors (Basel)*, *19*(1), 59. doi:10.339019010059 PMID:30586882

Zengshan, T., Jiacheng, W., Xiaolong, Y., & Mu, Z. (2018). WiCatch: A wi-fi based hand gesture recognition system access. *IEEE Access, 6*, 16911-16923.

Zhang, H., Xiao, Z., Wang, J., Li, F., & Szczerbicki, E. (2019). A novel IoT-perceptive human activity recognition (HAR) approach using multihead convolutional attention. *IEEE Internet of Things Journal*, *7*(2), 1072–1080. doi:10.1109/JIOT.2019.2949715

About the Contributors

Bhupesh Kumar Dewangan pursued Ph.D. in computer science and engineering from University of Petroleum and Energy Studies, Dehradun, India, and Master of Technology from Chhattisgarh Swami Vivekananda Technical University (State Technical University), Bhilai, India in computer science and engineering and Bachelor of Technology from Pandit Ravi Shankar Shukla University (State University), Raipur, India. He is currently working as an Associate Professor in the Department of Computer Science and Engineering, School of Engineering at the OP Jindal University Raigarh, India. He has more than 50 research publications in various international journals and conferences with SCI/SCOPUS/UGC indexing. He has three Indian patents on Cloud computing and resource scheduling. His research interests are in Autonomic Cloud Computing, Resource Scheduling, Software Engineering, and Testing, Image processing, and Object detection. He is a member of various organizations like ISTE, IAPFE, etc. Currently, he is editor in special issue journals of Inderscience & IGI publication house, and editor/author of two books of Springer & Taylor and Francis publication house.

* * *

Rinki Gupta received her Ph.D. in Signal Processing from the Centre for Applied Research in Electronics, Indian Institute of Technology Delhi, India in 2014. She joined Amity University in Nov. 2015. She is also the principal investigator in a Government funded research project and co-investigator in another. Her research interests include human motion analysis, machine learning and multi-sensor data fusion.

Princy Mishra is pursuing PhD in computer science and engineering and working as Assistant Professor (visiting) in department of computer science and engineering, at Amity University, Chhattisgarh, India, Bachelor of Technology from Chhattisgarh Swami Vivekananda Technical University (State Technical University), Bhilai in Information Technology and Master of Technology from Chhattisgarh Swami Vivekananda Technical University (State Technical University), Bhilai in computer science and engineering. Her research interest areas are Blockchain Technology, Machine learning and Image Processing.

Labdhi Sheth is currently a final year student pursuing her Bachelor's in Technology under the branch of Computer Science and Engineering from the Institute of Technology, Nirma University, Ahmedabad India. She is passionate about Digital Image and Video Processing and has worked on a couple of projects during her coursework. In 2021, she appeared as a guest lecturer for the webinar on Image processing. She is passionate about her work and wishes to explore more.

Bhavana Siddineni is currently pursuing B.Tech in Computer Science and Engineering at SRM University-AP, Andhra Pradesh, India. Her research interests include applying machine learning techniques in day-to-day life.

Mangal Singh is currently working as Associate Professor in the Department of Electronics & Telecommunication Engineering at Symbiosis Institute of Technology, Pune. He has an experience of more than 20 years in the field of Teaching, Research and Administration. He obtained his graduation in Electronics and Telecommunication Engineering from National Institute of Technology (formally known as GEC), Raipur, Chhattisgarh and M Tech in Communication Engineering Jadavpur University, Kolkata, West Bengal in 2000 and 2006, respectively. Dr Singh obtained his Ph D in Communication Engineering from National Institute of Technology, Rourkela, Odisha, in 2017. He has served as Associate Professor, Electronics & Communication Engineering, Institute of Technology, Nirma University, Ahmedabad from August 2018 to September 2021 and Associate Professor, Electronics & Communication Engineering, Chhatrapati Shivaji Institute of Technology, Durg, Chhattisgarh from September 2001 to July 2018. He has published more than 10 research papers in the area signal processing for communications, particularly multi-carrier modulation (OFDM) for wireless communication systems in International refereed/peer reviewed Journals and presented/published more than 20 papers in National/ International Conferences/Proceedings. He has 3 Indian patents published in his credit. He has guided more than 05 PG dissertations. He is a Senior Member of IEEE and life member of the IETE and ISTE, India.

Ishan Tewari is a final year student pursuing his Bachelor's in Technology under the branch of Computer Science and Engineering from the Institute of Technology, Nirma University, Ahmedabad India. He is passionate about Image Processing, Machine Learning, Deep Learning and iOS Development. He has worked on a lot of ML and DL Projects and has been invited to speak on ML at a webinar.

Shamik Tiwari, currently working as Sr. Associate Professor in School of Computer Science, University of Petroleum and Energy Studies, Dehradun. He has rich experience of around eighteen year as an academician and researcher. His area of interest are digital image processing, computer vision, bio-metrics, predictive modelling, statistical modelling, machine learning, deep learning and health informatics. He has written many national and international publications including books in these fields. He has published around 60 research papers in reputed journals and conferences, 4 patents and 5 books. He associated with many professional bodies and organizations as consultant, mentor, guide, and associate member. He is also serving as 'Adjunct Faculty' for Data Science and 'Thesis Supervisor' of upGrad Master's Program in collaboration with Liverpool John Moores University, UK.

Manikandan V. M. is currently working as an Asst. Professor in Computer Science and Engineering at SRM University-AP, Andhra Pradesh, India. He did his Ph.D. in Computer Engineering from the Indian Institute of Information Technology Design and Manufacturing Kancheepuram, Chennai, Tamilnadu, India after his M.Tech in Software Engineering from Cochin University of Science and Technology, Kerala, India. He is an Associate Member of The Institution of Engineers (India). His research interests include reversible data hiding, digital watermarking, digital image forensics, design and implementation of digital image processing algorithms, etc.

Index

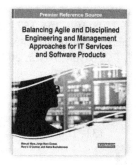

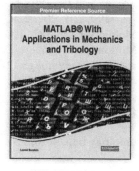

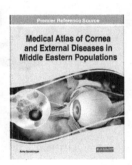

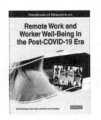